Information and Communication for Engineers

Other Macmillan titles of related interest

B. Allen, *Analogue Electronics for Higher Studies*
W.A. Atherton, *From Compass to Computer*
C.W. Davidson, *Transmission Lines for Communications, second edition*
M.E. Goodge, *Analog Electronics*
B.A. Gregory, *An Introduction to Electrical Instrumentation and Measurement Systems, second edition*
Robin Holland, *Microcomputer Fault-finding and Design, second edition*
Paul A. Lynn, *An Introduction to the Analysis and Processing of Signals, third edition*
R.J. Mitchell, *Microprocessor Systems – An Introduction*
Noel M. Morris, *Electrical Circuit Analysis and Design*
M.S. Nixon, *Introductory Digital Design*
R.G. Powell, *Electromagnetism*
R.C. Seals and G.F. Whapsholt, *Programmable Logic: PLDs and FPGAs*
P. Silvester, *Electric Circuits*
Andrew Simmonds, *Data Communications and Transmission Principles*
T.J. Terrell and Lik-Kwan Shark, *Digital Signal Processing*
M.J. Usher and D.A. Keating, *Sensors and Transducers – Characteristics, Applications, Instrumentation, Interfacing, second edition*
L.A.A. Warnes, *Electronic and Electrical Engineering – Principles and Practice*
L.A.A. Warnes, *Electronic Materials*
B.W. Williams, *Power Electronics – Devices, Drivers, Applications and Passive Components, second edition*

New Electronics Series

G.J. Awcock and R. Thomas, *Applied Image Processing*
Rodney F.W. Coates, *Underwater Acoustic Systems*
M.D. Edwards, *Automatic Logic Synthesis Techniques for Digital Systems*
Peter J. Fish, *Electronic Noise and Low Noise Design*
W. Forsythe and R.M. Goodall, *Digital Control*
C.G. Guy, *Data Communications for Engineers*
Paul A. Lynn, *Digital Signals, Processors and Noise*
Paul A. Lynn, *Radar Systems*
R.C.V. Macario, *Cellular Radio – Principles and Design, second edition*
A.F. Murray and H.M. Reekie, *Integrated Circuit Design*
F.J. Owens, *Signal Processing of Speech*
Dennis N. Pim, *Television and Teletext*
M. Richharia, *Satellite Communications Systems – Design Principles*
Peter Shepherd, *Integrated Circuit Design, Fabrication and Test*
M.J.N. Sibley, *Optical Communications, second edition*
P.M. Taylor, *Robotic Control*
G.S. Virk, *Digital Computer Control Systems*

Information and Communication for Engineers

M.J. Usher and C.G. Guy

University of Reading

MACMILLAN

First published 1997 by
MACMILLAN PRESS LTD
Houndmills, Basingstoke, Hampshire RG21 6XS
and London
Companies and representatives
throughout the world

ISBN 0–333–61527–1

A catalogue record for this book is available
from the British Library.

This book is printed on paper suitable for recycling and
made from fully managed and sustained forest sources.

10 9   8   7   6   5   4   3   2   1
06 05 04 03 02 01 00 99 98 97

Printed in Hong Kong

Contents

Preface

This book is based on an earlier text first published in 1984 and entitled: *Information Theory for Information Technologists*. That text was intended to provide a clear and thorough account of the principles of Information Theory in a manner suitable for students studying first degree courses in electronic engineering, cybernetics, control engineering, and telecommunications. Although the title was appropriate at the time, 'information technology' has since become a much 'softer' subject that is now studied by many students with more limited backgrounds in mathematics and physics; for this reason some sections of the book (for example, Chapter 2 and the appendices) may be unsuitable for such persons. However, the general treatment of information and its quantification (Chapter 1) and coding (Chapter 3) are relatively non-mathematical and should therefore provide a useful introduction for a wide class of readers. Since nowadays the subjects of information and communication cannot sensibly be separated, the book's title *Information and Communication for Engineers* both underlines this and reflects the intended readership.

In a book of this content the basic ideas about quantity of information must clearly come first, and four chapters from the original text have been combined into the new Chapter 1 entitled 'Information and its Quantification'. Some of the elementary mathematics on probability theory now appears in an appendix to avoid any disturbance to the flow of the book caused by a sudden excursion into mathematics with which some readers may already be familiar.

When information is to be transmitted, it is always in the form of some type of signal, so Chapter 2 is entitled 'Signals'. As it stands, it would provide useful introductory reading for a stand-alone course on signal theory. There is a substantial section on Fourier Theory, but Fourier Series are covered in an appendix. As explained in the book, Fourier Series are useful as an introduction to Fourier Theory, but once Fourier Transforms have been mastered, they become totally redundant. There are sections on the discrete and fast transforms, together with a short section on oversampling and noise shaping. Discussion of electrical noise is also included in an appendix.

It is explained in the book that when a signal is to be transmitted it is rarely in a suitable form and has to be modified in some way. Chapter 3 is entitled 'Coding' and includes discussion of Shannon's first and second theorems for coding. It provides a comprehensive treatment of the topic, including error detection and correction, and data compression techniques for text, speech and images.

Having encoded the information the next step is, of course, to transmit it and Chapter 4 is entitled 'Communication'. Again, a comprehensive treatment is provided, with sections on digital modulation, line codes and data communications.

The final chapter is entitled 'Applications'. This includes discussion of topics such as telephone and TV channels, as well as Teletext, FAX, CDs and computer networks. The aim is to show how some of the basic ideas of Information Theory have now found application in modern technology.

The book contains many worked examples together with examples for solution in each chapter, and with worked solutions at the end of the book.

1997   *M.J. Usher*
                                                          C.G. Guy

1 Information and its Quantification

1.1 Introduction

1.1.1 Information and Communication

Information Theory is concerned with the quantification and coding of information, and Communication Theory with its transmission. Some of the basic ideas were formulated by Nyquist (1924) and by Hartley (1928), but the subject received a considerable boost during and shortly after the Second World War, largely due to Shannon, who is regarded as the founder of the modern subject (see Shannon and Weaver, 1949). Some of Shannon's work was at first considered rather academic and impractical, but the technological developments of the last few years have changed this. Technology has caught up with information and communication, and modern developments are having great effects on our society. They may even make the writers redundant.

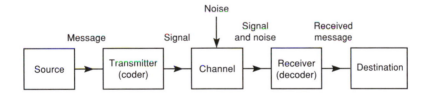

Figure 1.1.1 *A general communication channel*

In order to visualise the various processes involved in the communication of information we will consider the general communication system shown in Figure 1.1.1. This could represent a radio communication system in which music is being broadcast. The source would be a compact disc or tape but the music produced would have to be changed into suitable electrical signals (that is, 'coded') before it could be transmitted over the radio channel. Some noise, electrical interference etc. would be added in transmission, and more at the receiver. The receiver has to 'decode' the electrical signals and produce the music via a loudspeaker in order to be heard by the listener (the destination). Alternatively, the figure could represent a lecturer talking to his students; the source would be in his mind (or in his notes if he had not prepared the lecture), the coder would be the process of changing thoughts

1

into sound waves, and the noise would be the students chattering or shuffling their feet etc. The received message would in general differ somewhat from the transmitted message.

The essential parts of the information and communication process can thus be appreciated. We need to be able to analyse sources of information, to determine in particular the amount of information and its form. In general the information will not be in a suitable form for direct transmission in the channel, and various methods of coding it must be studied both in the sense of actual binary codes and of modulating some carrier signal. We will have to consider communication channels and their properties; in addition, since the information is usually transmitted in the form of an electrical signal, the study of signals and noise is very important. The decoding process may be similar to that in encoding, though different aspects are involved since we may be trying to determine the transmitted message from a seriously distorted signal. Overall, we may be interested in the accuracy of transmission, the speed of transmission and, of course, the amount of information transmitted.

1.1.2   Quantity of Information

Information has both qualitative and quantitative aspects. The two television programmes, *Panorama* and *Gladiators*, may have the same quantity of information in a mathematical sense, but are usually considered to have very different qualitative values. Sad to say, the latter probably has a greater quantity of information, being somewhat more spectacular, though the form of picture providing most information would be a totally random one (perhaps even more sleep-inducing than *Panorama* but certainly preferable to *Gladiators*).

Information Theory is concerned only with quantitative values, of course, and the first step in the subject is to define what we mean by quantity of information. A dictionary definition of information is 'Knowledge; intelligence given', which is not particularly helpful since the same dictionary defines both knowledge and intelligence in terms of information. However, in order to fix our ideas we will imagine that one person is speaking to another by telephone, imparting information, knowledge or intelligence. If the speaker says 'the sun has risen today' or 'your name is Mr Brown', the listener may be somewhat uninterested (unless his name is Mr White). However, if he says 'I am pleased to tell you that you have just won the lottery' the listener's reaction may be very different.

The amount of information conveyed in an event (a statement in this case) depends on the probability of the event. If someone is told something he already knows, so that the probability before being told it was already unity, the probability remains at unity after being told it. On the other hand, if one is told something that was relatively improbable, the probability changes from a small value before being told to unity afterwards. Similarly, if the speaker reads out letters of the alphabet, each letter is worth more information than if he reads out numerals (between 0 and 9) and considerably more than if he reads out a string of 0s and 1s.

A satisfactory definition of quantity of information must thus involve the probability of the event. Moreover, since longer messages convey proportionally more information than shorter ones, we must also ensure that our definition incorporates this. Another factor of importance is the reliability of the information; if we were unsure of the motives of the speaker in the examples above then even after the event the probability may be less than unity. However, in this chapter we will ignore this possibility, and our speaker will always tell the truth. We will consider a specific example in which the speaker is telling the listener the position of a piece on a chessboard. We will assume that a numbering system has been agreed in advance, as in Figure 1.1.2.

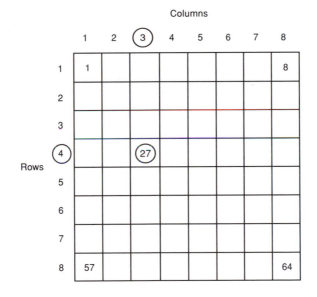

Figure 1.1.2 *Chessboard*

The speaker could give the position in two ways. He could simply state the number of the square, or he could separately give the column and the row. In the first case the probability before the event is 1/64 and afterwards unity. In the second case, when being told the row, the probability is 1/8 before and unity (for the row) afterwards; similarly it changes from 1/8 to unity for the columns. Clearly both methods must impart the same information; the sum of the information regarding the row (denoted I_a below) and that regarding the column (denoted I_b) must give that regarding the actual square (denoted I). We can see that the information is summed while the probabilities are multiplied ($1/8 \times 1/8 = 1/64$) and this immediately points to a definition involving a logarithm of the probability

$$\text{Quantity of information } I = -\log p$$

(a minus sign is chosen to make I positive, since $p \leq 1$). We can see immediately that this satisfies the above requirements.

(i) Number of square $I = -\log 1/64 = \log 64$
(ii) (a) Row $I_a = -\log 1/8 = \log 8$
       (b) Column $I_b = -\log 1/8 = \log 8$

Taking logs to base 2 for convenience, $\log_2 64 = 6$ and $\log_2 8 = 3$, so we have, as required $I = I_a + I_b$.

The definition thus satisfies our intuitive requirements; it depends on probability and messages add in the correct manner. If we take logs to base 2 the unit of information is known as the bit (binary digit). In a binary system producing 0s and 1s with equal probability, $p(0) = p(1) = 1$, so the information per digit is $\log_2 2 = 1$ bit. Other bases may be used, with correspondingly different units; using base e for example, one gets nats (*natural units*), using base 3 (with some care) one gets trits, and base 10 produces 'Hartleys' after one of the original workers who evidently became more interested in fly fishing. However, since base 2 is almost always used, we will assume in future that the base is 2 (unless specified otherwise) and omit the subscript.

We will illustrate the definition further with a few examples.

(i) *Letters of the alphabet.* If the letters are assumed equiprobable, then the probability of a given letter is 1/26 and $I = \log 26 = 4.7$ bits.
(ii) *Numerals.* Assuming that numbers from 0 to 9 are equiprobable, $I = \log 10 = 3.32$ bits.
(iii) *An already-known event.* If $p = 1$, then $I = \log 1 = 0$.
(iv) *Sequence of binary digits.* Suppose the sequence 1 0 1 1 of four digits is sent. If 0s and 1s are equiprobable then the information received is 4 bits. Alternatively, the probability of obtaining this sequence is $(1/2)^4 = 1/16$, so the information is $\log 16 = 4$ bits again. This confirms again that the definition gives the correct result regarding the length of the message. However, if the digits are not equiprobable the information would be less than 4 bits (see below).

A different approach to defining quantity of information is to say that the information in an event must be proportional to the amount of recording medium required to store it, for example in recording speech on tape. Considering a simple case of storing numerals ranging from 0 to 999, they can obviously be stored as three decimal digits but would require ten binary digits ($2^{10} = 1024$ so numbers up to 1024 could be stored in 10 binary digits). Similarly, numbers between 0 and 10^6 require six decimal digits or twenty binary digits. Clearly the number of digits needed to store numbers between 0 and N is $\log_{10}N$ in decimal or $\log_2 N$ in binary. If the numbers are equiprobable, $p = 1/N$ and we obtain the same definition $I = -\log p$. However, this method fails to emphasise the inherent importance of probability in quantifying information and cannot be applied if the numbers are not equiprobable.

1.1.3 Average Information: Entropy

In practice we are often more interested in the average information conveyed in some process than in the specific information in each event. For example, a source may produce a set of events of probability $p_1, p_2, p_3, \ldots$ $p_i \ldots p_n$. In a long sequence of n events, event 1 will occur $n \times p_1$ times contributing $n \times -\log p_1$ bits, event 2 will occur $n \times p_2$ times contributing $n \times -\log p_2$ etc. The average information over all the events is called the entropy H, and is given by

$$\text{Entropy } H = -\sum_i p_i \log_i \text{ bits}$$

In the case of letters of the alphabet, the probabilities are not really all the same (as assumed above) and the values for English text are shown in Table 1.1.1.

Table 1.1.1 *Letter probabilities in English text*

Letter	Probability	Letter	Probability	Letter	Probability	Letter	Probability
a	0.0703	h	0.0451	o	0.0565	v	0.0077
b	0.0104	i	0.0571	p	0.0198	w	0.0165
c	0.0255	j	0.0019	q	0.0010	x	0.0020
d	0.0317	k	0.0050	r	0.0507	y	0.0148
e	0.1010	l	0.0278	s	0.0540	z	0.0006
f	0.0202	m	0.0229	t	0.0821	space	0.1820
g	0.0132	n	0.0561	u	0.0241		

The entropy is given by

$$
\begin{aligned}
H &= -(p(\text{A}) \log p(\text{A}) + p(\text{B}) \log p(\text{B}) + p(\text{C}) \log p(\text{C}) \ldots \\
&\quad + p(\text{Z}) \log p(\text{Z})) \\
&= -(0.0703 \log(0.0703) + 0.0104 \log(0.0104) \\
&\quad + 0.0255 \log(0.0255) \ldots + 0.0006 \log(0.0006)) \\
&= -(0.0703 \times 3.830 + 0.0104 \times 6.587 + 0.0255 \times 5.293 \ldots \\
&\quad + 0.0006 \times 10.70) \\
&= -(-0.269 - 0.069 - 0.135 \ldots - 0.064) \\
&= 4.1 \text{ bits (using the values in Table 1.1.1)}
\end{aligned}
$$

It can be seen that although a Z provides much more information than an A (because it is less probable), it occurs so rarely that the average information due to Zs is much less than the average due to As. The units are often given as bits/letter or bits/symbol to stress the average nature of the measure.

The term 'entropy' was deliberately chosen for the name of the measure of average information, because of its similarity to entropy in thermodynamics. In thermodynamics, entropy is a measure of the degree of disorder of a system, and disorder is clearly related to information. The paradox of Maxwell's demon, who separates hot and cold molecules and thereby contravenes the second law of thermodynamics, can be solved (somewhat philosophically) by assuming that the demon requires 'information' in order to effect the separation.

EXAMPLE 1.1.1

A binary source produces a stream of 0s and 1s with probabilities $p(0) = 1/8$ and $p(1) = 7/8$ respectively. Find the entropy.

$$H = -\Sigma\, p_i \log p_i = -(1/8 \log 1/8 + 7/8 \log 7/8) = 0.55 \text{ bits}$$

It is very important to realise that a binary digit is not the same as a bit of information. A stream of binary digits can carry a maximum of one bit per digit, which occurs when the 0s and 1s are equally probable. The term binit is sometimes used to mean binary digit, stressing its difference from the bit. A plot of H versus $p(0)$, as $p(0)$ varies from 0 to 1, is shown in Figure 1.1.3. The entropy is unity only for $p(0) = p(1) = 1/2$, falling to zero for $p(0) = 1$ or $p(0) = 0$. This is because a continuous stream of digits of the same type (all 1s or all 0s) does not convey any information. In a

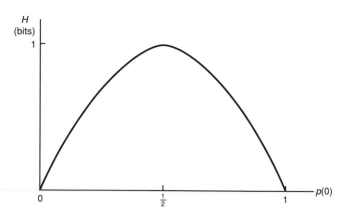

Figure 1.1.3 *Variation of average information with probability for a binary source*

similar way, the entropy of a set of events is a maximum when they are equiprobable, as in the case considered above for letters of the alphabet.

1.1.4 Redundancy

Redundancy is an important concept in Information Theory, particularly in connection with language. It is the presence of more symbols in a message than is strictly necessary. For example, in a binary system of two symbols A and B, we could code A as 0 0 0 and B as 1 1 1 (instead of A = 0, B = 1). This gives some protection against errors, since one binary error in three digits could be tolerated (assuming a majority is used for decision at the receiver). For example: 1 0 0, 0 1 0 and 0 0 1 would be decoded as A; 0 1 1, 1 0 1 and 1 1 0 would be decoded as B; but two or three binary errors would produce a symbol error.

Redundancy $R = 1 - $ (actual entropy)/(maximum entropy)

In the above example, $R = 1 - 1/3 = 2/3$, since three digits could carry three bits, but actually carry only one. Spoken languages usually have high redundancy, permitting them to be understood in the presence of noise or errors. For example, if one received the message

CAMINK HINE TINOGHT

most people would easily understand what was intended. English has a redundancy of about 80 per cent, and this will be considered in some detail in section 1.2.

1.1.5 Examples for Solution

These exercises can be done without a calculator, given that $\log_2 3 = 1.58$ and $\log_2 5 = 2.32$. However, it is generally much quicker to use one, and the reader is strongly advised to do so. The formula for entropy lends itself readily to such evaluation. The preferred method is to work in logs to base 10, and to multiply by 3.322 (that is, $\log_2 10 = 1/\log_{10} 2$) at the end.

1. What is the gain in information on being told that when a die was thrown:
    (i) a six occurred
    (ii) a six or a one occurred
    (iii) a one, two, three, four, five or six occurred?
2. A picture is composed of 20 rows and 20 columns of dots, each dot having three possible degrees of darkness. Calculate the information in the picture:
    (i) assuming that the shades are equally probable
    (ii) assuming that the probabilities are 1/8, 1/4, 5/8.
    Note: You may assume that it is agreed in advance how the dots will be arranged (as is usually the case in transmitting pictures).

3. A source produces two symbols A and B with probabilities 1/4 and 3/4 respectively. Find the average information received per second:
  (i) assuming that A and B each take 1 s
  (ii) assuming that A takes 1 s and B takes 2 s.

1.2   Information in Language

1.2.1   Conditional Entropy

As we have seen above, the information per symbol of a set of N independent and equiprobable symbols is simply log N, giving a value of 4.7 bits per symbol for the 26 letters of English. Taking account of the probabilities of individual letters the average information, or entropy, is given by $H = -\Sigma\, p_i \log p_i$, producing a lower value of about 4.1 bits/symbol.

Unfortunately the practical situation is more complicated because the letters are not independent either. In a sequence of letters of the alphabet the probability of, say, a U after a Q is very high whereas that of a Q after a U is quite low. Similarly, in many sequences of letters there is strong dependence of one letter on another or on several successive letters. The effect is known as intersymbol influence and the probabilities that describe them as conditional probabilities. It is interesting to note that the effect is the equivalent of the bandwidth of a communication channel, which governs the rate at which the signal may change.

As discussed in Appendix 1, the joint probability of two independent events A and B, of probabilities $p(A)$ and $p(B)$, occurring together is given by

$$p(AB) = p(A) \times p(B)$$

For two conditionally dependent events A and B, the expression for the joint probability changes to

$$p(AB) = p(A) \times p(B/A)$$

where $p(B/A)$ means the probability of B given that A has already occurred. For example, in English text the value of $p(TH) = p(T) \times p(H/T) = 0.0255$ so $p(H/T) = p(TH)/p(T) = 0.0255/0.0821 = 0.311$. This is very much greater than the probability of an independent H (0.0451) so the information provided by the H is much reduced by the intersymbol influence.

Consider a sequence of letters in which intersymbol influence extends only over pairs of adjacent letters. For two such letters i and j, the information obtained when j is received is $-\log p(j/i)$; that is, minus the logarithm of the probability of receiving j, given that i has been received. In order to find the average information over all letters we simply have to average over all possible pairs of letters ij. The result is known as the conditional entropy $H(j/i)$ of the sequence, and represents the information per letter in such a sequence:

$$H(j/i) = -\sum_i \sum_j p(ij) \log p(j/i)$$

EXAMPLE 1.2.1

A simple language consists of only two symbols A and B produced in a long continuous sequence. Find the single, joint and conditional probabilities of A and B, assuming that the values found from the limited sequence below are typical of a very long sequence (assume that the 21st letter is A in order to have 20 pairs of symbols). Evaluate the conditional entropy for the sequence, and hence deduce the redundancy of the language.

AABBBAAAABBAAABBBAAA

$p(A) = 12/20$, $p(B) = 8/20$ by simply counting As and Bs.
$p(AA) = 9/20$, $p(BB) = 5/20$, $p(AB) = 3/20$, $p(BA) = 3/20$ by counting pairs.
$p(A/B) = 3/8$, $p(B/B) = 5/8$ by counting the number of times an A or B comes after a B
$p(A/A) = 9/12$, $p(B/A) = 3/12$ by counting As or Bs after an A.

There are four pairs of symbols AA, BB, AB and BA. Therefore

$$\begin{aligned}
H(j/i) &= -(p(A\ A) \log p(A/A) + p(B\ B) \log p(B/B) \\
&\quad + p(A\ B) \log p(B/A) + p(B\ A) \log p(A/B)) \\
&= 9/20 \log 9/12 + 5/20 \log 5/8 + 3/20 \log 3/12 \\
&\quad + 3/20 \log 3/8 \\
&= 0.868 \text{ bits/symbol}
\end{aligned}$$

If no intersymbol influence had been present the information would have been given by

$$\begin{aligned}
H(i) &= -\sum p(i) \log p(i) = -(p(A) \log p(A) + p(B) \log p(B)) \\
&= -(0.6 \log 0.6 + 0.4 \log 0.4) \\
&= 0.971 \text{ bits/symbol}
\end{aligned}$$

The redundancy is given by $R = 1 -$ actual entropy/maximum entropy. The maximum entropy would occur for independent and equiprobable symbols and is clearly 1 bit per symbol, so $R = 1 - 0.868/1 = 13$ per cent.

It can be seen that most of the redundancy is due to the intersymbol influence between A and B, the effect of symbol probabilities only reducing the information from 1.0 bits/symbol to 0.97 bits/symbol. The effect is even more pronounced in a practical language of more than the two symbols chosen here.

If intersymbol influence extends over several symbols, so that conditional probabilities of the form $p(k/ij)$ arise (that is, the probability of a k after the sequence i, j), the formulae are of the same general form. For example, $H(k/ij) = -\sum\sum\sum p(i, j, k) \log p(k/ij)$ etc.

1.2.2   Redundancy in Printed English

The redundancy in language arises in several ways. The letters have widely different probabilities, as shown in Table 1.1.1, and strong intersymbol influence exists not only over adjacent symbols but also over whole groups of symbols. Suppose we imagine that a word is being spelled out to us in a telephone conversation. There is a choice of 26 letters (though not equally probable) for the first letter, providing about 4 bits when received; suppose it was a D. The choices for the next letter are restricted, because of the construction of English, to about 7 letters, providing about 2.8 bits; assume it was an I. There is a greater choice next, of all letters, giving another 4 bits. However, if the next few letters were S, T, R the choice rapidly diminishes and after a further I, B, U the word can be reasonably guessed to be say DISTRIBUTION or DISTRIBUTIVE. The last few letters clearly provide very little information, because the probability of the next letters after the sequence DISTRIBU approaches unity.

Similarly, words are not equiprobable and intersymbol influence extends over groups of words, phrases, sentences etc., and the result is that the information content of language is very much less than the ideal 4.7 bits/symbol of a language of 26 equiprobable and independent symbols. It is likely that an optimum amount of redundancy is automatically achieved in the development of a language, such that it can be sufficiently understood in the presence of noise and errors. In some special cases the language does not have sufficient redundancy, and extra redundancy must be added. The NATO Phonetic Alphabet (alpha, bravo etc.) was designed for use when accurate spoken data are to be transmitted by radio or telephone, and is widely used by the military, police and even in telephone banking services.

Shannon and Weaver (1949) studied the redundancy of printed English in a particularly interesting and instructive manner, by constructing successively closer approximations to complete text. Their zeroth approximation was a language consisting of 26 letters and a space symbol, all equiprobable, providing $\log_2 27 = 4.75$ bits/symbol. The space is actually the most probable symbol, so a sample of text from such a source looks wrong even from a distance (when the letters cannot be read).

The first approximation consisted of putting in the correct letter probabilities; the information can then be calculated from $H = -\Sigma p_i \log p_i$ and comes to 4.1 bits/symbol. Such text looks correct from a distance, but is unpronounceable.

The second approximation involves using conditional probabilities over pairs of adjacent symbols, calculating the information from the formula $H(j/i) = -\Sigma p(i, j) \log p(j/i)$, and yields a value of 3.3 bits/symbol. Tables of conditional probabilities are available, and it can be seen that a significant reduction of information occurs, the corresponding text being reasonably readable (though meaningless, of course). A third approximation, in which conditional probabilities over three symbols were considered, had a rather small effect, the information being calculated as 3.1 bits/symbol.

Further approximations involved using the correct probabilities for complete words (about 2.1 bits/symbol) and allowing for conditional probabilities between words. It becomes progressively more difficult to produce samples of text, and for the later approximations these were generated by picking letters or words out of a book (which automatically has the correct probabilities built into it).

Finally, in order to estimate the information in complete English text, 'volunteers' were used who were shown samples of text and asked to predict the next letters. This is a rather lengthy process but it was possible to show that the information varies between 0.6 and 1.3 bits/letter depending on the type of text. Presumably in some detective stories the content is rather low (since one can guess the outcome) whereas it is hoped that this book is at the higher end of the range!

A different approach was used by Bell (1953), who considered complete words (assumed equiprobable). Bell estimated the numbers of words of various lengths from a dictionary, and produced a table similar to that of Table 1.2.1.

Table 1.2.1

Number of letters	1	2	3	4	5	6	7	8	> 8
Number of words	2	26	722	2166	3610	4765	4260	3861	8447
Entropy/letter	1.0	2.4	3.2	2.8	2.3	2.0	1.7	1.5	0.7

Finding 26 two-letter words would tax the imagination of most people (except players of a certain unmentionable word game) but, if equiprobable, each word provides log 26 = 4.7 bits or 2.35 bits/symbol. The information per letter rises with word length, reaching a maximum for three and four letter words. Apparently expletives are very efficient at conveying information! The content then falls rapidly, as expected from the discussion above. An average over all word lengths gives 2.6 bits/letter, which is rather higher than Shannon's figure because of the assumption of independent and equiprobable words.

It thus appears that reasonable average values for the information content of English are about 2 bits/letter for complete words and about one bit per letter for text. The redundancy of text is therefore about 1 − 1/4.7 = 80 per cent. As we saw above, this is probably an optimum for a spoken language, though a lower value may be appropriate for a written language. In particular, it can be seen that the storage of text in computers is very inefficient. Text is usually stored in ASCII form, in which each letter is stored as a 7-bit code. An 8-bit 'byte' is usually used for convenience, so nearly 8 times as much space is used as is ideally necessary. However, as will be shown in Chapter 3, it is impractical to employ a code that provides a saving of more than about 50 per cent.

1.2.3 Examples for Solution

1. A binary source produces a sequence of Marks and Spaces such that $p(M/S) = 1/3$, $p(S/S) = 2/3$, $p(M/M) = 5/6$, $p(S/M) = 1/6$.
   (a) Find $p(M)$, $p(S)$, $p(MM)$, $p(SS)$, $p(MS)$, $p(SM)$.
   (b) Find the conditional entropy and the redundancy.
2. A source produces three symbols A, B and C. After B the probability of a B is 1/2, of A 1/4 and of C 1/4. After A the probability of A is 5/8, of B 1/4 and of C 1/8. After C the probability of C is 5/8, of B 1/4, and of A 1/8. Find the redundancy of the system.
3. A primitive language consists of five symbols designated a, b, c, e and Ω, grouped into words of one, two and three letters. Ω is used only as the space symbol between words. The only one-letter words are a and e. Two-letter words begin or end with a or e and three-letter words always have a or e as the middle letter (and nowhere else). Find the total number of words and hence find the information per symbol and redundancy, assuming all words to be equally probable and independent.

   Discuss, with reasons, whether the language is likely to be spoken or written (or both).
4. Assume that the sequence below is part of a long sequence of a newly discovered language, comprising only two symbols A and B:

   AAABABAABBAAAABAABAA

   (a) Use the sequence to find the single, joint and conditional probabilities of the two symbols, assuming that conditional probabilities extend only over adjacent symbols.
   (b) Hence find the average information per symbol and the redundancy of the language.
   (c) How would you determine whether conditional probabilities extend over more than two symbols, and how would the redundancy be calculated if they do?
   (d) Is the language likely to be spoken or written?
   (Assume that the 21st symbol is A, in order to obtain 20 pairs.)

1.3 Information in Noisy Channels

1.3.1 Noisy Channels

We saw in section 1.1.2 that the quantity of information received from an event of probability p was given by $-\log p$ or $\log 1/p$. The probability of the event at the receiver changed from p before the event to unity after the event. However, in many practical situations the probability after the event will be a little less than unity, because there may be some uncertainty about the validity of the transmission. For example, if numerical data are read over a telephone line the listener may be unsure that he heard some of the values correctly, or in the transmission of data by binary pulses some of the pulses may be received in error.

Random Noise

Random noise occurs in all physical processes. In order for a physical system to attain thermal equilibrium with its surroundings it must exchange energy, and the result is a fluctuation in the energy of the system which gives rise to random noise in some form. (This is a statement of the Principle of Equipartition, which is discussed further in Appendix 2.) The fluctuations appear as Brownian motion in mechanical systems, as changes in temperature in thermal systems and as Johnson noise in electrical systems. Johnson noise occurs in any electrical component that dissipates power, so it occurs in resistors, in capacitors and inductors (because they have a non-zero ohmic resistance) and in transistors (because of their base resistance). Other types of electrical noise such as Shot noise (due to the discrete nature of electrical charge) and $1/f$ noise also occur in most semiconductor devices. It is clear that the amplifiers and receivers used in practical communication systems will be subject to random electrical noise, leading to the possibility of errors in the data transmitted.

Binary data may be transmitted as a succession of positive and negative electrical voltages as shown in Figure 1.3.1 on page 14. A '1' is represented by a positive level lasting one clock cycle and the level then remains the same for the next cycle if another '1' is transmitted, or falls to a negative value for a '0'. Such a signal is known as a random telegraph signal. The data are sent over the channel (not necessarily in this form) and 'reproduced' at the output of the channel. The receiver, which may be assumed to keep in step with the transmitter, must make a decision as to whether a '1' or a '0' is present. This is done by sampling the output at the centre of a clock cycle and decoding as '1' if the voltage is greater than zero or '0' if less than zero.

Errors may occur because of the noise added to the signal in the transmission and reception processes. A simple binary channel is characterised by a certain mean binary error rate, which depends on the signal power transmission and bandwidth used.

Quantity of Information in a Noisy Channel

Consider a hypothetical system consisting of a transmitter T, a receiver R and two channels, one of which is noisy and the other noiseless, as shown in Figure 1.3.2.

An event of probability p_0 is first transmitted over the noisy channel. At the receiver the probability after reception is p_1, a little less than unity because

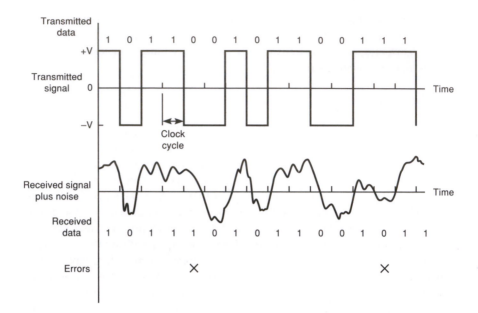

Figure 1.3.1 *Errors due to random noise in a binary channel*

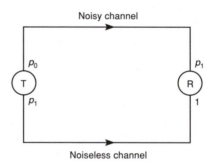

Figure 1.3.2 *Quantity of information in a noisy channel*

of the possibility of error. A second transmission of the same event is now made via the noiseless channel and the probability at the receiver changes from p_1 before reception to unity after reception, as the channel is noiseless. The total information transmitted is $I + \log (1/p_1)$, where I is the information in the noisy channel. Now the same information could have been obtained by a single transmission via the noiseless channel, the probabilities changing from p_0 to unity with information $+\log (1/p_0)$. Thus

$$I + \log (1/p_1) = \log (1/p_0)$$

$$\text{or } I = \log (p_1/p_0)$$

p_0 is known as the *a priori* probability, the probability at the receiver of the event before transmission. p_1 is known as the *a posteriori* probability, the probability at the receiver of the event after transmission. It can be seen that for a noiseless system p_1 would be unity and the expression would become $I = -\log p_0$ as before.

EXAMPLE 1.3.1

A binary system produces Marks and Spaces with equal probability, 1/8 of all pulses being received in error. Find the information received for all possible combinations of input and output.

There are four possibilities: M → M, M → S, S → M, S → S.

For the case M → M (and similarly S → S), the probability at the receiver that an M was transmitted after an M has been received is 7/8. Thus

$$I \text{ (M} \rightarrow \text{M or S} \rightarrow \text{S)} = \log (7/8)/(1/2) = 0.81 \text{ bits}$$

For the case M → S (and similarly for S → M), the probability at the receiver that an M was transmitted after an S has been received is 1/8. Thus

$$I \text{ (M} \rightarrow \text{S or S} \rightarrow \text{M)} = \log (1/8)/(1/2) = -2.0 \text{ bits}$$

The average information received is: $0.81 \times 7/8 + (-2.0) \times 1/8 = 0.46$ bits.

A number of important observations can be made:

(a) Information is reduced even when the transmission is actually correct. In a noiseless system the information would have been one bit, but the possibility of error has reduced it to 0.81 bits. This can be seen to be reasonable, by considering transactions on the stock market. If there is some doubt about the validity of data received, the investor will invest less than he might otherwise and therefore make less profit even if the data turn out to have been correct.

(b) An error leads to negative information. The receiver has no way of knowing which transmissions are in error, of course, so in the stock market example above the investor would lose if the data were wrong.

(c) The average information is always greater than (or equal to) zero. An error rate of 0.5 would make the average information zero. However, a greater error rate causes the information to increase again, because if every pulse was wrong the receiver would have totally correct data by a simple inversion.

This latter observation is greatly beloved of lecturers and authors, because it means that however bad their lecture or book it is almost certain to provide positive information on average! It also shows how politicians give nothing away; they lie exactly half the time but at random.

1.3.2 General Expression for Information Transfer

For a noiseless system we saw that the expression $I = -\log p$ for the quantity of information in a single event could be generalised to give the average information $H = -\Sigma\, p_i \log p_i$ over a set of events of probabilities p_i. We will do the same thing for a noisy system, denoting inputs by x and outputs by y. Our previous formulae for conditional probabilities apply directly here, because p_1 (the probability at the receiver after the event) is clearly such a probability.

Suppose we have a set of input symbols x_i and a corresponding set of outputs y_i (so that $x_i \to y_i$ in the absence of error). The *a posteriori* probability p_1 is given by $p(x_i/y_i)$, that is, having received a certain symbol y_i the probability that a certain symbol x_i was transmitted. For this event we have

$$I(x_iy_i) = \log\,[p(x_i/y_i)/p(x_i)]$$

Since $p(xy) = p(x)\,p(y/x) = p(y)\,p(x/y) = p(yx)$, (the order of events does not matter, unlike the sequences considered earlier)

$$I(x_iy_i) = \log\,[p(x_i/y_i)/p(x_i)] = \log\,[p(y_i/x_i)/p(y_i)]$$
$$= \log\,[p(x_iy_i)/p(x_i)\,p(y_i)]$$

To obtain the average information over all events we have only to multiply by the probability of the event $p(x_iy_i)$ and sum. The result is called the *Information Transfer or Mutual Information $I(xy)$*. In the expression below the suffixes i have been omitted for clarity and the summations are over all x or all y as shown.

$$I(xy) = \sum_x \sum_y p(xy) \log \frac{p(xy)}{p(x)p(y)}$$

(any of the previous three terms may be used, but this form is preferred).

This result is rather surprising in that it is symmetrical in x and y, that is, x and y are interchangeable! This gives us a rather different interpretation of information exchange; the information transferred is not unidirectional but is a measure of the degree of similarity between receiver and transmitter. In fact, the expression for information transfer is quite similar to the mathematical expression for the correlation coefficient between two sets of data.

Suppose one set up a system comprising a box with digital indicators (representing, say, some binary number) and a second similar box but arranged such that there were occasional random differences between the two numbers displayed. This could be a communication system with a certain error rate. A person entering the room would not know which was the receiver and which the transmitter, yet he could work out the information transfer (if he had read this book). Indeed, it would be physically unreasonable if he could tell which was which, as it would imply causality in random devices.

It can be seen that the general expression behaves satisfactorily, for if x and y are independent then $p(xy) = p(x) \times p(y)$ and the expression has value zero since $\log(1) = 0$. Similarly, if x and y are totally dependent (that is, in a noiseless system) $p(xy) \to p(x) = p(y)$ and $I \to -\Sigma p(x) \log p(x) = H(x)$.

EXAMPLE 1.3.2

A binary system produces Marks with probability 0.7 and Spaces with probability 0.3. 2/7 of the Marks are received in error and 1/3 of the Spaces. Find the information transfer.

This type of problem is best done by writing out a 'typical' sequence of M and S (that is, with the correct probabilities) and evaluating the required probabilities:

$$x \quad \text{M M M M M M M S S S}$$
$$y \quad \text{M M M M M S S S S M}$$

Table 1.3.1 shows the probabilities obtained by counting the required events. (Note that $p(x/y)$ and $p(y/x)$ are not needed here, but are included for completeness and because they will be required later.)

Instead of using this expression one could use one with $p(x/y)$ or $p(y/x)$, but it is easier to work out the joint probabilities $p(xy)$ than conditional ones. The error rate in this example is impractically high of course, and the error rates have been chosen to simplify computation. It can be seen that very little information is actually transmitted.

Table 1.3.1

x	M	M	S	S
y	M	S	S	M
$p(x)$	0.7	0.7	0.3	0.3
$p(y)$	0.6	0.4	0.4	0.6
$p(x/y)$	5/6	1/2	1/2	1/6
$p(y/x)$	5/7	2/7	2/3	1/3
$p(xy)$	0.5	0.2	0.2	0.1
$I(xy)$ terms	0.126	−0.997	0.147	−0.085

$$\begin{array}{cc} \text{MM} & \text{MS} \\ I(xy) = \; 0.5 \log(0.5)/(0.7\times0.6) \; + \; 0.2 \log(0.2)/(0.7\times0.4) \end{array}$$

$$\begin{array}{cc} \text{SS} & \text{SM} \\ + \; 0.2 \log(0.2)/(0.3\times0.4) \; + \; 0.1 \log(0.1)/(0.3 \times 0.6) \end{array}$$

$$= 0.091 \text{ bits/symbol}$$

1.3.3   Equivocation

The idea of 'equivocation' provides another interpretation of information transfer. It represents the destructive effect of noise on information, or alternatively (which is the same thing) the additional information that the receiver requires in order to correct all his data. Consider a source and receiver again, connected by both a noisy and a noiseless channel as before, but this time with an 'observer' using the noiseless channel as in Figure 1.3.3.

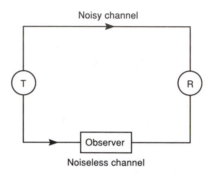

Figure 1.3.3   *Noisy channel with a hypothetical observer*

The observer looks at each pair of transmitted and received digits; if they are the same he sends the receiver a '1', if different a '0'. A sequence of transmissions could be as follows:

x	M	M	S	S	M	M	M	S	M
y	M	S	S	S	M	M	M	S	S
observer	1	0	1	1	1	1	1	1	0

The information sent by the observer is easily evaluated. It is simply $-(p(0) \log(p0) + p(1) \log p(1))$ applied to the binary stream sent. Moreover, since he only sends a '0' for an error, the probability of '0' is just the channel error probability p_e.

EXAMPLE 1.3.3

Find the information sent by the (hypothetical) observer for a binary system with equiprobable inputs and an error rate of 1/8 (example 1.3.1 again).

The observer sends $7/8 \log (7/8) + 1/8 \log (1/8) = 0.55$ bits. Since the input information $H(x) = 1$ bit per symbol, the net information (without the observer) is $1 - 0.55 = 0.45$ bits, agreeing with the previous result. Alternatively we can say that noise in the system has destroyed 0.55 bits of information or that the equivocation is 0.55 bits.

General Expression for Equivocation

Considering a specific pair of transmitted and received digits, the probability at the receiver changes from $p(x)$ to $p(x/y)$ after reception down the noisy channel. After receiving the observer's correcting data the probability changes to unity; that is the observer provides $-\log p(x/y)$.

Averaging over all pairs as before, we obtain a general expression for equivocation $H(x/y)$:

$$H(x/y) = -\sum_x \sum_y p(xy) \log p(x/y)$$

The information transferred via the noisy channel (in the absence of the observer) must be

$$I(xy) = H(x) - H(x/y)$$

This expression can be deduced directly from our previous expressions for information transfer (since it clearly must agree):

$$
\begin{aligned}
I(xy) &= \Sigma\Sigma\, p(xy) \log p(xy)/p(x)p(y) \\
&= \Sigma\Sigma\, p(xy) \log p(x/y)/p(x) \quad \text{(since } p(xy) = p(y)\, p(x/y)) \\
&= \Sigma\Sigma\, p(xy) \log p(x/y) - \Sigma\Sigma\, p(xy) \log p(x) \\
&= -H(x/y) + H(x) \quad \text{(since } \Sigma_y\, p(xy) = p(x))
\end{aligned}
$$

A similar derivation gives $I(xy) = H(y) - H(y/x)$ in which $H(y/x)$ is a 'backwards' equivocation.

EXAMPLE 1.3.4

Solve example 1.3.2 using the expression for equivocation.

In this case we have to work out the conditional probabilities. We need only either $p(x/y)$ or $p(y/x)$ but both will be used for completeness. Referring again to Table 1.3.1, in which $p(x/y)$ and $p(y/x)$ are evaluated:

$$
\begin{aligned}
H(x) &= -\Sigma\, p(x) \log p(x) = -\,(0.7 \log 0.7 + 0.3 \log 0.3) \\
&= 0.881 \text{ bits}
\end{aligned}
$$

$$
\begin{aligned}
H(y) &= -\Sigma\, p(y) \log p(y) = -\,(0.6 \log 0.6 + 0.4 \log 0.4) \\
&= 0.970 \text{ bits}
\end{aligned}
$$

$$
\begin{aligned}
H(x/y) &= -\Sigma\Sigma\, p(xy) \log p(x/y) \\
&= 0.5 \log 5/6 + 0.2 \log 1/2 + 0.2 \log 1/2 + 0.1 \log 1/6 \\
&= 0.790 \text{ bits}
\end{aligned}
$$

$$
\begin{aligned}
H(y/x) &= -\Sigma\Sigma\, p(xy) \log p(y/x) \\
&= 0.5 \log 5/7 + 0.2 \log 2/7 + 0.2 \log 2/3 + 0.1 \log 1/3 \\
&= 0.879 \text{ bits}
\end{aligned}
$$

Thus

$$
\begin{aligned}
I(xy) &= H(x) - H(x/y) = 0.091 \text{ bits} \\
&= H(y) - H(y/x) = 0.091 \text{ bits}
\end{aligned}
$$

This agrees with the value found previously using the formula for information transfer. For completeness we can also calculate the joint entropy $H(xy)$ given by

$$H(xy) = -\Sigma\Sigma\, p(xy)\, \log p(xy)$$
$$= 0.5\, \log 0.5 + 0.2\, \log 0.2 + 0.2\, \log 0.2 + 0.1\, \log 0.1$$
$$= 1.76 \text{ bits}$$

It is easily proved that $H(xy) = H(x/y) + H(y/x) + I(xy)$ and the values above confirm this. However, the concept of joint entropy is not very important as it does not have any useful interpretation.

1.3.4 Summary of Basic Formulae by Venn Diagram

The various concepts and formulae discussed in this chapter, and the relations between them, can be usefully summarised in a Venn diagram. If we imagine that all possible information is represented by an unbounded plane, we can represent the information at the transmitter of a specific system by (say) a circle in the plane. The inside of the circle is the source entropy $H(x)$. If the system was noiseless the entropy $H(y)$ at the receiver would be represented by an identical circle coincident with that for $H(x)$. However, in a noisy system the two circles would intersect only partially and if x and y were independent they would not intersect at all. The area of intersection thus represents the information transfer $I(xy)$. Similarly, the equivocations $H(x/y)$ and $H(y/x)$ are represented by the remainder of the $H(x)$ and $H(y)$ circles respectively when $I(xy)$ is removed, and the joint entropy $H(xy)$ is the area bounded by the perimeter of the figure. The relations between the various measures can easily be seen (or remembered) from Figure 1.3.4.

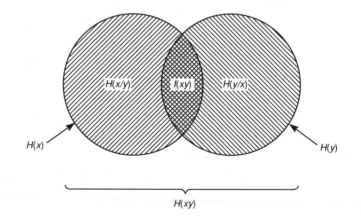

Figure 1.3.4 *Venn diagram summarising basic formulae*

Source entropy: $\quad H(x) = -\Sigma\, p(x)\, \log\, p(x)$

Receiver entropy: $\quad H(y) = -\Sigma\, p(y)\, \log\, p(y)$

Equivocations: $\quad H(x/y) = -\Sigma\Sigma\, p(xy)\, \log\, p(x/y)$

$\qquad\qquad\qquad\quad H(y/x) = -\Sigma\Sigma\, p(xy)\, \log\, p(y/x)$

Information transfer: $I(xy) = +\Sigma\Sigma\, p(xy)\, \log\, p(xy)/p(x)\, p(y)$

$\qquad\qquad\qquad\qquad\quad = H(x) - H(x/y)$

$\qquad\qquad\qquad\qquad\quad = H(y) - H(y/x)$

Joint entropy: $\quad H(xy) = -\Sigma\Sigma\, p(xy)\, \log\, p(xy)$

$\qquad\qquad\qquad\qquad\quad = H(x/y) + H(y/x) + I(x/y)$

1.3.5 *Channel Capacity*

An information channel will normally have a fixed known bandwidth and a fixed known (or measurable) error probability. The *capacity* C of the channel is defined as the maximum information transfer, with respect to the probabilities of the input symbols (which are the only variables), that is

$\quad C = \max\, I(xy)$

The relations between input and output in a general channel can be conveniently represented in matrix form, the elements of the matrix giving the conditional probabilities $(p(x/y)$ or $p(y/x)$ etc.). One can then calculate the capacity of a specific channel by finding the distribution of input probabilities that gives the maximum transfer. This would be very tedious in most cases. However, binary channels are by far the most common type, and of these the binary symmetric channel is the most usual. We will therefore concentrate only on this type of channel.

Binary Symmetric Channels

Referring again to the binary system shown in Figure 1.3.1, the decision level at the receiver, which determines whether the received digit is a '0' or a '1', is usually chosen at 0 V for a bipolar system. If the noise in the system is random the probabilities of errors in '0' and '1' will be the same (a different choice of decision level could clearly give more errors in 0s or 1s). Such a channel is known as a binary symmetric channel and is characterised by a single value p of binary error probability. The possible transitions from transmitter to receiver are shown in Figure 1.3.5.

The capacity of this channel has a very simple form and can be deduced as follows:

$I(xy) \quad = H(y) - H(y/x)$

$\qquad\quad = H(y) + \Sigma\Sigma\, p(xy)\, \log\, p(y/x)$

$\qquad\quad = H(y) + \Sigma_x\, p(\mathrm{x})\, (\Sigma_y\, p(y/x)\, \log\, p(y/x))$

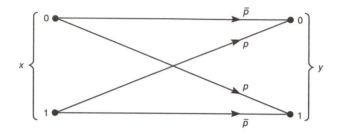

Figure 1.3.5 *Binary symmetric channel*

Considering the summation over y, for a given x one of the values of y represents an error and the other represents a correct transmission, so the two probabilities are just p for an error and $\bar{p}$ $(= 1 - p)$ for no error. Therefore

$$I(xy) \;=\; H(y) + \Sigma_x \, p(x) \, (p \log p + \bar{p} \log \bar{p})$$

$$=\; H(y) - H(p) \quad \text{where } H(p) = -\,(p \log p + \bar{p} \log \bar{p})$$

$H(p)$ is an equivocation and its formula is just like that for the entropy of a noiseless binary system with probabilities p and $\bar{p}$.

Now $I(xy)$ will be a maximum when $H(y)$ is a maximum, since p and therefore $H(p)$ are fixed. This will occur when $p(0) = p(1)$ at the receiver, and since the system is symmetric, when $p(0) = p(1)$ at the transmitter. The channel capacity is therefore

$$C = 1 - H(p)$$

and is obtained for equiprobable input symbols.

Figure 1.3.6(a) shows how $I(xy)$ varies with probability of input digits, reaching a maximum (when it becomes the channel capacity) of $1 - H(p)$ for $p(0) = p(1)$. Figure 1.3.6(b) shows how the capacity C varies with binary error probability p, being unity for $p = 0$ or $p = 1$ and zero for $p = 1/2$.

EXAMPLE 1.3.5

Find the capacity of a binary symmetric channel with a binary error rate of 0.125.

This is a thinly disguised example 1.3.1 yet again.

$$H(p) \;=\; -\,(p \log p + \bar{p} \log \bar{p}) = -\,(1/8 \log 1/8 + 7/8 \log 7/8)$$
$$=\; 0.55 \text{ bits}$$

$$C \;=\; 0.45 \text{ bits}$$

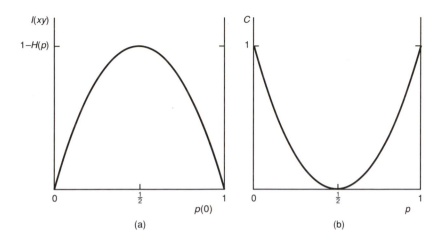

Figure 1.3.6 *(a) Variation of information transfer with input probability. (b) Variation of capacity with error probability*

1.3.6 Examples for Solution

1. Symbols A, B and C occur with equal probability. Two-thirds of As are received as A, one-sixth as B and one-sixth as C. C behaves in a corresponding manner and no errors occur in B. Find the information transfer.
2. A binary symmetric channel has a binary error probability of 1/3. Two equiprobable input symbols A and B are to be transmitted, coded as 11 and 00 respectively. Determine:
   (a) the channel capacity
   (b) the information transfer through the channel, assuming that the receiver interprets 11 as A, 00 as B and makes no decision for 01 or 10
   (c) the information transfer if the receiver interprets 11, 01 or 10 as A, and 00 as B
   (d) the information transfer if the receiver interprets 01 and 10 randomly as A or B.
   Explain the reasons for the different values in (b), (c) and (d).
   Note: The probabilities of A being received as 11, 01, 10 or 00 are respectively 4/9, 2/9, 2/9 and 1/9.
3. Two binary symmetric channels are to be cascaded. The first has a binary error probability of 0.02 and the second of 0.04. Assuming that the input symbols are equiprobable find the information transfer:
   (a) through the first channel alone
   (b) through the second channel alone
   (c) through both channels in cascade
   Comment on the general effect of cascading noisy channels.

4. An information source produces three symbols A, B and C with probabilities
1/4, 1/4 and 1/2 respectively. Errors occur in transmission as shown below. Find the probabilities of A, B and C at the receiver and hence deduce the information transfer.

Input	Output		
	A	B	C
A	2/3	1/3	0
B	0	1	0
C	0	1/3	2/3

1.4   Information in Continuous Signals

1.4.1   Continuous Signals

So far we have restricted our attention to discrete signals, such as letters of
the alphabet or streams of binary digits. However, many practical signals
are essentially analogue (or continuous, so that they have an unlimited resolution
as opposed to the discrete steps in a digital signal), such as a speech signal
from a microphone or a radio signal, and we need to be able to deduce their
information capacity. A typical analogue signal is shown in Figure 1.4.1.

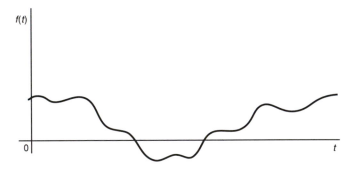

Figure 1.4.1   *Continuous analogue signal*

Such signals are often digitised in practice, so that they may be processed by computer. The rule for such digitising is provided by the sampling
theorem, which is discussed in detail in Chapter 2. The sampling theorem
states that if a signal is limited in bandwidth to *B* Hz then provided it is
sampled at a rate of at least 2*B* samples per second, all the information in
the analogue signal is retained and the samples can be converted back to
analogue with no loss in information content. A speech signal of bandwidth
3 kHz must therefore be sampled at a rate of at least 6000 per second,

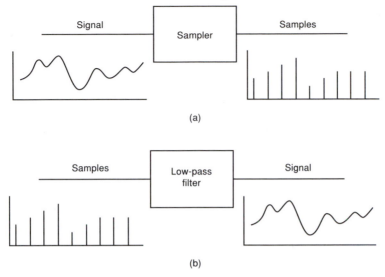

Figure 1.4.2 *Sampling (a) and recovery (b) of an analogue signal*

though in practice a higher rate (typically 8000 per second) is usually used to make reconversion to analogue form easier. Figure 1.4.2 shows the sampling and recovery process.

The sampling theorem provides a simple way of estimating the information content of an analogue signal. If we digitise the signal, the information per second is given by:

R = (number of samples/s) × (information per sample)

There are a number of possible objections to this equation. Firstly we can, of course, take as many samples as we wish, not just $2B$/s. What is really required is the maximum number of *independent* samples/s. Fortunately this is just the value provided by the sampling theorem. If we take more than $2B$/s the waveform will be easily reproducible from the samples, but the samples will not be independent. In contrast, if we take fewer than $2B$/s the waveform will not be reproducible but the samples will be independent. The critical number $2B$/s is the value required; this rate is also referred to as the Nyquist signalling rate.

The second objection is more serious. From our previous definition of information in discrete signals as the logarithm of the probability of an event, the information per sample can be taken as the logarithm of the number of distinguishable levels (assuming for the present that all levels are equiprobable). However, if the signal is not perturbed by noise the number of possible levels is infinite. In principle, if we have a sufficiently good meter we can measure the amplitude of a given sample as accurately as we wish, so there is no limit to the number of levels, even if the maximum amplitude is fixed. The information per sample is therefore infinite! This is unfortunate. One intuitively feels that different types of continuous signal should have different information capacity, a random signal being capable of carrying more information than, say, a mainly sinusoidal signal. However, apparently they do all have the same capacity – infinity!

The solution lies in recognising that any practical system will be perturbed by random noise, and that this will limit the number of distinguishable levels. If the root mean square noise level is σ, a r.m.s. meter will read σ for zero signal. It is found practically that the smallest detectable signal has amplitude about σ (that is, root mean square signal-to-noise ratio unity), the next about 2σ etc., so if the maximum signal amplitude is s the number of distinguishable levels is about s/σ. We thus obtain an expression for the information per second, usually identified with the channel capacity C, given by:

$$C = 2B \times \log (s/\sigma)$$

This is often rewritten in terms of signal-power-to-noise-ratio S/N, usually denoted SNR. Writing signal power $S = s^2$ and noise power $N = \sigma^2$ we have $\text{SNR} = s^2/\sigma^2$ and

$$C = 2B \log (\text{SNR})^{1/2} = B \log (\text{SNR})$$

Our intuitive approach has thus yielded a simple yet specific result, almost identical to the result of a more thorough analysis which we shall now consider.

1.4.2 Relative Entropy of Continuous Signals

We have shown previously that the entropy of a discrete set of events of probability p_i is given by $H = -\Sigma p_i \log p_i$. When dealing with continuous signals, discrete probabilities must be replaced by probability densities $p(v)$. Figure 1.4.3 shows a probability density distribution for the signal of Figure 1.4.2.

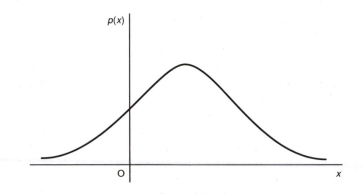

Figure 1.4.3 *Probability density distribution*

The probability of the signal being in the range $v_1 < v < v_2$ is given by $\int_{v_1}^{v_2} p(v)dv$. The area under the graph of $p(v)$ is always unity because the probability of the signal having a value between zero and infinity is $\int_{-\infty}^{\infty} p(v)dv$ and must clearly be unity. The expression for the entropy of a continuous signal is therefore obtained by replacing p_i in the discrete case by $p(v)dv$ and replacing the summation by an integral, and is given by:

$$H(v) = -\int_{-\infty}^{\infty} p(v) \log p(v)dv$$

It represents the information per sample of a continuous waveform, and depends on $p(v)$ as we intuitively require. Its use will not cause any difficulties provided we remember that all practical signals are perturbed by noise, but the expression must not be used blindly in isolation; it is strictly known as the *relative entropy*.

The type of signal having greatest information capacity will clearly have a probability density distribution $p(v)$ which maximises $H(v)$. One might expect this to be a uniform distribution (since H for a discrete system is greatest when all the probabilities are equal), but a better guess would be a Gaussian distribution (since there is no limit to amplitude and it applies to random noise). There are mathematical techniques for obtaining the maximising distribution, which does indeed turn out to be the Gaussian, but they will not be reproduced here as they add nothing (except about a page of working) to the subject. We will therefore simply evaluate $H(v)$ for a Gaussian distribution for which $p(v)$ is given by:

$$p(v) = \frac{1}{\sigma \sqrt{2\pi}} \exp \left(\frac{-v^2}{2\sigma^2} \right)$$

The integral is easier to evaluate if we work in natural units (nats), taking logs to base e:

$$H(v) = -\int_{-\infty}^{\infty} p(v) \log_e p(v)dv \text{ nats}$$

$$= \int_{-\infty}^{\infty} p(v) \log_e (\sigma\sqrt{2\pi}) \, dv + \int_{-\infty}^{\infty} p(v) \frac{v^2}{2\sigma^2} \, dv$$

$$= \log_e(\sigma\sqrt{2\pi}) + 1/2$$

$$= \log_e(\sigma\sqrt{2\pi e}) \text{ nats (since } 1/2 = \log_e(e^{0.5}))$$

We can now change the base back to 2 and the units simply change to bits! Thus

$$H(v) = \log_2 (\sigma\sqrt{2\pi e}) \text{ bits}$$

Writing power $P = \sigma^2$, the final expression becomes

$$H(v) = \log(2\pi e P)^{0.5} \text{ bits}$$

This derivation is much admired by those who are peculiarly appreciative of any results that can be obtained without doing any real work!

EXAMPLE 1.4.1

Compare $H(v)$ for the following waveforms, normalising such that all waveforms have a variance of unity: (i) Gaussian distributed, (ii) repetitive triangular, (iii) square.

(i) The Gaussian gives $\log(\sigma(2\pi e)^{1/2})$ bits as shown above, giving 2.05 bits for $\sigma = 1$.
(ii) Figure 1.4.4(a) and (b) shows a typical triangular wave and its probability density distribution.

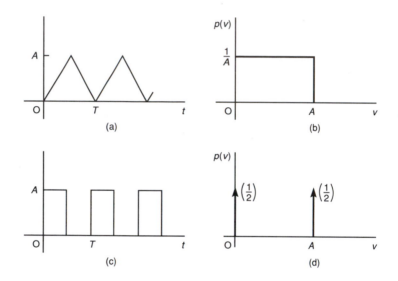

Figure 1.4.4 *Triangular wave (a) and its probability density distribution (b); square wave (c) and its probability density distribution (d)*

$$\overline{v^2} = \int_0^A v^2\, p(v)\mathrm{d}v = A^2/3, \qquad \overline{v} = A/2, \qquad \sigma^2 = A^2/12$$

$$H(v) = -\int_0^A \frac{1}{A}\, \log\, \frac{1}{A}\, \mathrm{d}v = \log A = \log(\sigma\sqrt{(12)}),\ \text{giving 1.79}$$

bits with $\sigma = 1$

    (the same value will be found for any triangular wave.)
(iii) Figure 1.4.4(c) and (d) shows a square wave and its probability density distribution. $p(v)$ consists of two impulses, each of strength one-half, at O and A.

$$H(v) = -\int_{-\infty}^{\infty} p(v)\, \log p(v)\mathrm{d}v = -(1/2\, \log 1/2 + 1/2\, \log 1/2)$$

$$= 1\ \text{bit}$$

(since the only contributions to the integral are at $v = 0$ and $v = A$, and $p(v)$ is then $1/2$).

Note that this is the same result we get for a discrete system with two levels, and that the height of the levels does not matter. The ratios of $H(v)$ for the three cases are $1:0.87:0.49$, showing that the Gaussian distribution carries more information than the uniform distribution, as expected. The term 'entropy power' is sometimes used in comparing $H(v)$ for different waveforms.

1.4.3 Information Capacity of Continuous Signals

We want our signal to be as general as possible, so we will place as few restrictions on it as we can. We will consider a non-repetitive signal of mean power S perturbed by noise of mean power N and limited in bandwidth to the range 0 to B Hz (any bandwidth B is actually sufficient). We will assume that both the signal and noise are Gaussian distributed, since we require the maximum information. These are the only specifications required; in fact only the signal-power-to-noise ratio (SNR) is needed, not S and N separately.

Figure 1.4.5 shows a simple communication system. The signal power at the input x and output y are both S (assuming that the gain in the system is suitably adjusted) but noise power N is present at the output. The observed output has power $S + N$, since there is no way of distinguishing between signal and noise (they are both Gaussian distributed).

Figure 1.4.5 *Information transfer in a continuous channel*

From our previous formula the information transfer $I(xy)$ is

$$I(xy) = H(y) - H(y/x)$$

The equivocation $H(y/x)$ represents the effect of noise, since x and y would be identical in a noiseless system, so we can replace it by $H(n)$, meaning the noise entropy, where $H(n) = \log(2\pi eN)^{1/2}$ from above. Now $H(y) = \log(2\pi e(S + N))^{1/2}$ since the output power is $(S + N)$. The capacity C is therefore $H(y) - H(n) = \log((S + N)/N)^{1/2}$. Assuming that we take $2B$ samples/s as before, the capacity (expressed as information/s) is given by

$$C = B \log (1 + SNR) \text{ bits/s} \tag{1.4.1}$$

Note that this is almost identical with the result obtained by our simple intuitive approach in section 1.4.1; in fact, we could have obtained an identical result by taking the number of levels to be $((s/\sigma) + 1)$.

This is a very important and significant expression. It refers to a very general signal, yet tells us the information per second in terms of only the bandwidth B and the signal-power-to-noise ratio (SNR). Although it may not be apparent from the above derivation, it has a further significance. Not only is it the maximum information transfer – it is also the maximum error-free communication rate! Equation 1.4.1 is known as the *ideal communication theorem* and the capacity as the ideal communication rate. Information can be transmitted at any rate up to and including C with no net errors at all.

A formal proof will not be given here, but we will outline a method given by Shannon, since it is very instructive and gives another insight into the situation.

Suppose we have a signal lasting for a time T so that the information carried is $C \times T$ or $BT \log (1 + \text{SNR})$ bits, and that the signal is sampled at the Nyquist rate giving $2BT$ samples. We can (attempt to) imagine our signal (of $2BT$ values) to be represented by a single point in a multi-dimensional space of $2BT$ dimensions! Another signal, of the same characteristics and duration, would be represented by another point in the same space. The signal will be perturbed by noise, whose effect will be to 'blur' the signal so that instead of being represented by a point it becomes a small 'sphere' centred on the point. The number of distinguishable signals is just the number of such spheres that can be packed into the space. The properties of multi-dimensional spaces provide the required result, and it is found that if the number of samples tends to infinity the above expression for C is indeed the maximum error-free rate! The actual derivation is somewhat lengthy (see Rosie, 1973, pp. 96–103).

Of course, the objection can again be raised that the expression or its derivation does not tell us how to go about setting up a system to communicate at or near this rate. However, at least we know that we must use Gaussian distributed waveforms of long duration, and schemes similar to that of Figure 1.4.6 have been suggested (Rice, 1950).

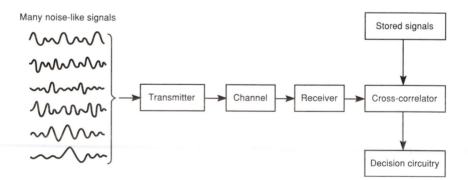

Figure 1.4.6 *Hypothetical 'ideal' communication system*

A large number of different signals represent the messages to be transmitted. They are all bursts of random noise of the same duration. The same signals are also stored at the receiver and when a particular signal (blurred by noise) is received it is cross-correlated with each of the stored signals in turn, and identified by the largest output. The system suffers from serious practical problems; it is necessary to use an impossibly large number of signals, to select them to be as different from one another as possible, and very long time delays occur in the decision process.

It is instructive to compare the capacity and error rate of a simple binary system with that given by the ideal theorem. Suppose a binary signal, such as that discussed previously in Figure 1.3.1, has an amplitude of ± 1 V and is perturbed by noise of root mean square amplitude 0.31 V in a bandwidth of 1 kHz (this corresponds to a signal-to-noise ratio of 3.16 and a signal-power-to-noise ratio of 10). It can be shown that the binary error rate is 8.0×10^{-4} (see Appendix 2). If the probabilities of 0s and 1s are equal then the capacity C is given by $C = 1 - H(p) = 0.99$ bits. Assuming that signalling is at the maximum rate of $2B$ binary pulses per second, the information rate will be $2000 \times 0.99 = 1981$ bits/s.

Ideally $C = B \log (1 + \text{SNR}) = 1000 \times \log_2 (11) = 3459$ bits/s with no net errors at all. The binary system is surprisingly good, considering that it is at first sight not much like random noise, though in fact a good approximation to such noise can be generated by low-pass filtering a random binary train of pulses.

1.4.4 Deductions from the Ideal Theorem

Although we cannot at present approach very close to the ideal communication rate predicted by equation 1.4.1, it serves as a useful standard against which to compare practical systems. In addition, several important deductions can be made which do have useful practical applications. The information transmitted in a time T is

$$I = BT \log(1 + \text{SNR})$$

as above. It is clear that a given amount of information I can be transmitted by many different combinations of bandwidth B, duration T and signal-power-to-noise ratio SNR. For example, keeping the bandwidth constant we can exchange SNR and T; a trivial example of this is simply repeating a message. Similarly B and T may be exchanged; for example, one could record a message on tape at high speed (large B, small T) and play back slowly (small B, large T) down, say, a telephone channel. These two exchanges are obvious of course, and do not owe much to the ideal theorem; however, the exchange between B and SNR is much less obvious, yet very important.

We will use numerical values to illustrate the possibilities, and let the information I have a value of 3 bits. Figure 1.4.7 is the locus of values of B and SNR for constant I (= 3 bits) and T (= 1 second).

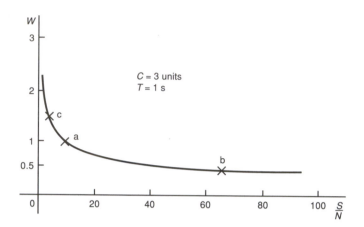

Figure 1.4.7   *Exchange of bandwidth and signal-power-to-noise ratio*

We could try to save on B by increasing SNR, or to save on SNR by increasing B. Such possibilities are very important in practice. There is great pressure on bandwidth in broadcast systems whereas signal power is less critical, so a saving in bandwidth could be very valuable. However, the opposite requirements apply in communications with a space probe; the power on the probe is strictly limited but there may be adequate bandwidth allocated.

We will take the point a in Figure 1.4.7, where $B = 1$ and SNR $= 7$, as a reference. If we halve the bandwidth the corresponding SNR must be 63 (point b), whereas if we halve SNR the corresponding bandwidth becomes about 1.5 units (point c). The first possibility, reducing bandwidth by increasing power, is simply not on; it requires a vast increase in power of nine times in this case to halve the bandwidth. However, the second possibility is very much on; the power can be halved at the cost of only a 50 per cent increase in bandwidth!

The question now arises, of course, as to how the exchanges could be carried out. Considering the impractical one first, two successive samples of a waveform could be combined into one larger value which could be transmitted in a narrower bandwidth (that is, at half the sampling rate). This is illustrated in Figure 1.4.8, where the input waveform is quantised to one of eight possible levels.

The binary codes for two successive samples of 6 units (110) and 7 units (111) are combined into the code 110111 ($= 55$), and a single pulse of 55 units transmitted (in a 64-level system). The power is increased from $(6^2 + 7^2)$ to $(55)^2$, or by a factor of about 35, though the required bandwidth is halved.

The second possibility arises naturally in several typical communication systems, such as frequency modulation (FM) and pulse-code modulation (PCM). In these systems the signal spectrum is modified such that the transmission

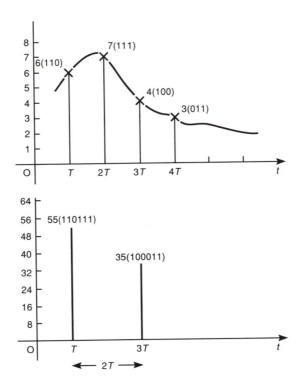

Figure 1.4.8 *Reduction of bandwidth by combining samples*

bandwidth on the channel is much larger than the original signal bandwidth, and the result is that much less transmitter power is required. The technique of PCM is rather similar to that shown in Figure 1.4.8. The signal is quantised into one of a given number of levels, but instead of two samples being combined into one large pulse a train of binary pulses is transmitted which represents the samples in binary (for example, 110111 for the first two samples in Figure 1.4.8). The bandwidth is increased by a factor of 3 (in this case), and the required power considerably reduced. Quantisation/sampling will be described in more detail in Chapter 2, and PCM and FM in Chapter 4.

Another interesting deduction is of the minimum power required to transmit one bit of information. The total noise power N can be written as $N_0 \times B$, where N_0 is the power spectral density of the noise. The maximum value of C occurs for B tending to infinity and SNR tending to zero; that is

$$C_{max} = \lim [B \log (1 + S/N_0 B) \text{ bits}$$

Working in natural units, since $\lim \log_e(1 + x) = x$, we have

$$C_{max} = BS/N_0 B = S/N_0 \text{ nats}$$

The value of N_0 for Johnson noise is kT, so we have the surprising result that the minimum amount of power required to transmit one natural unit of information is kT Watts (about 4×10^{-21} W). Usually one works in bits of course (small ones where students are concerned), but it could be said that the nat is a bit more natural!

1.4.5 Examples for Solution

1. Find the relative entropy of the waveform of Figure 1.4.9 and compare it with that of Gaussian noise, normalising to unity mean power in each case.

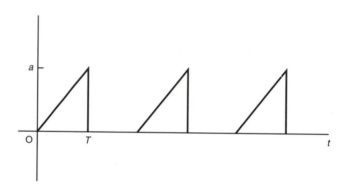

Figure 1.4.9 *Sawtooth waveform*

2. A telephone channel has a bandwidth of 3 kHz and a signal-power-to-noise ratio of 100. Compare the rate of transmission of information when data are transmitted in the form of binary pulses with that for an ideal communication system of the same parameters. *Note*: assume that the error rate for the binary pulses is zero.

3. A television signal has a bandwidth of 8 MHz and a signal-power-to-noise ratio of 1000. The signal is displayed on a monochrome TV screen which can be considered to consist of a square matrix of 625 × 625 pixels, each of one of 32 equiprobable levels of brightness, with 25 complete pictures transmitted per second. Estimate the information content in the picture and hence the information rate, and compare with the value deduced from the ideal theorem. Why are errors less important in this system than in the telephone system of question 2?

2   Signals

2.1   Introduction

A dictionary definition of 'signal' is 'a transmitted effect conveying a message'. We have so far discussed only the quantification of information, but when we come to transmit the information it will always be in the form of some type of signal, such as the amplitude-modulated sine wave or the sampled speech waveform of Figure 2.1.1. This chapter is concerned with the properties and characteristics of signals, with emphasis on the information and communication aspects.

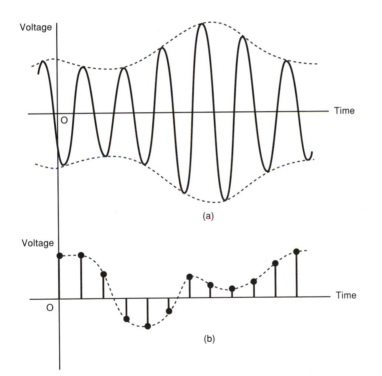

Figure 2.1.1 *(a) Amplitude-modulated sine wave; (b) sampled speech signal*

An essential characteristic of a signal is that of change, since it must be capable of carrying information. We have seen that a stream of digits that are all ones carries no information, and similarly a constant quantity cannot carry information. The change must be at least partly unpredictable, otherwise of course there would be no information. However, it is necessary also to discuss the properties of sinusoidal or pulse waveforms because, although they are totally predictable, they can be modulated to carry information as in Figure 2.1.1.

Signals may be functions of time (temporal signals), as in the electrical signal in a telephone line, or functions of space (spatial signals), as in the case of a newspaper picture. They may be both spatial and temporal, as in a television picture, but the important element in each case is that of change. Most of the signals that we have to deal with will be temporal and, although the equations for both temporal and spatial signals are identical in form, we will use the notation $f(t)$ for all such signals.

2.1.1 Finite Energy and Finite Power Signals

An important classification of signals is into 'finite energy' and 'finite power' signals. The instantaneous power P_{inst} in a signal $f(t)$ is defined by:

$$P_{inst} = f^2(t)$$

The signal will usually represent either voltage or current, and the units will then be mean square volts or mean square amperes. In Signal Theory such quantities are always collectively referred to as 'power', although the units are usually not Watts.

The energy in a signal in a small time interval δt is $f^2(t) \, \delta t$, so the total energy E over all time is given by

$$E = = \int_{-\infty}^{\infty} f^2(t) \mathrm{d}t$$

For example, the rectangular pulse of height A and width θ of Figure 2.1.2 has instantaneous power A^2 (from $t = 0$ to $t = \theta$) and energy $A^2\theta$.

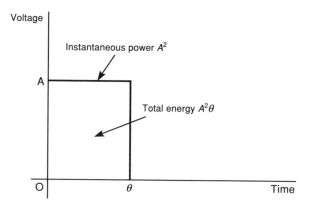

Figure 2.1.2 *Energy in a rectangular pulse*

In contrast, the repetitive pulses of Figure 2.1.3 have infinite energy, since each pulse contributes $A^2\theta$ and they are without limit. However, the pulses have finite mean power ($A^2\theta/T$) since power is energy per unit time, whereas the single pulse of Figure 2.1.2 has zero mean power. Mean power is formally defined by:

$$P = \lim_{T\to\infty} \frac{1}{T} \int_{-T/2}^{T/2} f^2(t)\mathrm{d}t = \frac{1}{T} \int_{-T/2}^{T/2} f^2(t)\mathrm{d}t \quad \text{if repetitive with}$$

period T

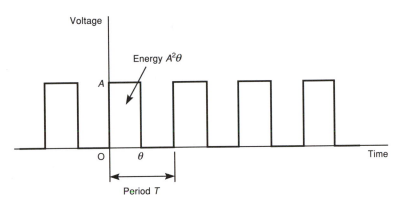

Figure 2.1.3 *Power and energy in an infinite train of pulses*

It can be seen that the description of signals in terms of energy or power depends on their form. The most important forms are as follows:

(a) *Transient signals*: these have finite energy and zero mean power. They usually last for a finite time, such as the rectangular pulse of Figure 2.1.2, but an exception is an exponentially decaying waveform which in a sense lasts for all time but nevertheless has finite energy.
(b) *Continuous signals*: these have finite mean power but infinite energy and last for all time. They may be either *repetitive*, with a constant repetition period T as for the train of pulses of Figure 2.1.3, or *random*, as in the case of analogue noise or a random pulse train.

Signals may be analysed in the time domain, by considering various types of time average, or in the frequency domain, by considering the distribution of power or energy with frequency.

2.2 Basic Time and Frequency Averages

2.2.1 Time Domain Averages

Consider the continuous signal $f(t)$ shown in Figure 2.2.1.

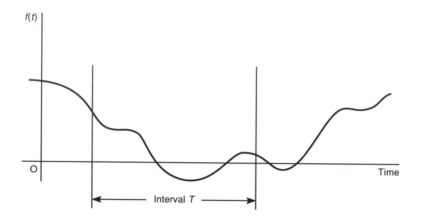

Figure 2.2.1 *Mean values for a continuous signal*

The *average* of $f(t)$ over the interval T is defined as

$$\text{average} \qquad \bar{f}_T = \frac{1}{T} \int_{-T/2}^{T/2} f(t)\,\mathrm{d}t$$

and the time average (over all time), known as the *mean value* is

$$\text{mean value} \qquad \bar{f} = \lim_{T\to\infty} \frac{1}{T} \int_{-T/2}^{T/2} f(t)\,\mathrm{d}t$$

$$\left(= \frac{1}{T} \int_{-T/2}^{T/2} f(t)\,\mathrm{d}t \quad \text{if repetitive with period } T \right)$$

The *mean square value* is defined by:

$$\text{mean square value } \overline{f^2} = \lim_{T\to\infty} \frac{1}{T} \int_{-T/2}^{T/2} f^2(t)\,\mathrm{d}t$$

$$\left(= \frac{1}{T} \int_{-T/2}^{T/2} f^2(t)\,\mathrm{d}t \quad \text{if repetitive} \right)$$

This is, of course, the same as the mean power defined above.

The *variance* is given by

$$\text{variance } \sigma^2 = \lim_{T \to \infty} \frac{1}{T} \int_{-T/2}^{T/2} \left(f(t) - \bar{f} \right)^2 dt$$

$$\left(= \frac{1}{T} \int_{-T/2}^{T/2} \left(f(t) - \bar{f} \right)^2 dt \text{ if repetitive} \right)$$

$f(t) - \bar{f}$ is the instantaneous difference between $f(t)$ and its mean value $\bar{f}$, so σ^2 represents the mean square deviation from the mean value. The square root of this is usually called the *standard deviation*. By expanding the integral it is easy to show that

$$\sigma^2 = \overline{f^2} - (\bar{f})^2$$

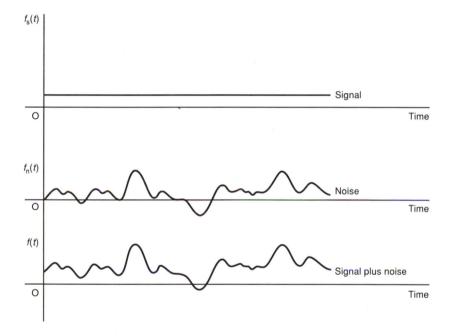

Figure 2.2.2 *Detection of steady signal in noise*

The relationship between these three quantities can be seen by considering the detection of a signal represented by a steady voltage in the presence of random noise, as in Figure 2.2.2.

If the noise is random it will have a zero mean value. The mean value of $f(t)$ is therefore the required signal. The mean square value gives the total power of signal and noise together, and the variance gives the noise power; that is

$$\text{total power} = \text{signal power} + \text{noise power}$$

$$\overline{f^2} \quad = \quad (\overline{f})^2 \quad + \quad \sigma^2$$

EXAMPLE 2.2.1

Find the mean value, mean square value and variance for a sine wave $f(t) = A \sin \omega_0 t$.
 The mean value $\overline{f}$ is clearly zero, so $\overline{f^2} = \sigma^2$.

$$\overline{f^2} = \frac{1}{T} \int_{-T/2}^{T/2} (A \sin \omega_0 t)^2 \, dt \text{ where } T = \frac{2\pi}{\omega_0}$$

$$= \frac{A^2}{T} \int_{-T/2}^{T/2} \left(\tfrac{1}{2} - \tfrac{1}{2} \cos 2\omega_0 t \right) dt = A^2/2$$

The above definitions apply only to finite power signals. In the case of finite energy signals the only applicable averages are the area of the signal and the total energy. For example, in the case of the transient signal of Figure 2.2.3 the area is $\int_{-\infty}^{\infty} f(t) dt$ and the total energy $\int_{-\infty}^{\infty} f^2(t) dt$, whereas for the pulse of Figure 2.1.2 the values are $A\theta$ and $A^2\theta$ respectively.

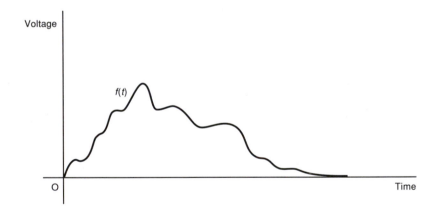

Figure 2.2.3 *Transient signal*

2.2.2 *Frequency Domain Averages*

(a) *Finite Power Signals*

Consider the continuous signal of Figure 2.2.1 again and suppose it is fed to a tuneable filter and mean square meter as in Figure 2.2.4 (most meters

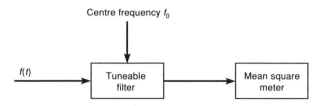

Figure 2.2.4 *Apparatus for measuring power spectrum*

measure the root mean square value, but here we require the square of this).

We will assume that the filter passes a narrow band of frequencies of bandwidth δf with a centre frequency f_0, which can be varied over a wide range without affecting δf, and that the gain is unity. For a continuous random signal the plot of meter reading versus centre frequency f_0 would be similar to Figure 2.2.5. If $f(t)$ represented a voltage the units would be mean square volts per bandwidth δf.

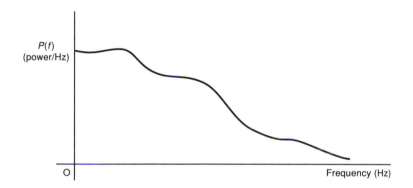

Figure 2.2.5 *Power spectrum of a continuous signal*

The power spectrum $P(f)$, sometimes called power spectral density, is the distribution of mean power per unit bandwidth and is obtained by dividing the mean square voltage by δf, so the units are mean square Volts/Hz or Watts per Hz. Clearly, the total mean power $\overline{f^2}$ (or P) is given by

$$P = \int_0^\infty P(f)\mathrm{d}f,$$ that is, the integral of $P(f)$ over all frequencies.

In the case of a repetitive signal, such as in Figure 2.1.3, the power spectrum is found to consist of discrete lines spaced at multiples of the repetition period T, as in Figure 2.2.6.

The line at zero frequency is simply the square of the mean value, that is, $(A\theta/T)^2$, and is known as the zeroth harmonic. That at frequency $1/T$ is the fundamental or first harmonic, and that at $2/T$ is the second harmonic

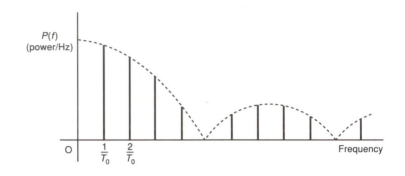

Figure 2.2.6 *Power spectrum of a repetitive signal*

etc. (Many people confuse the first and second harmonics, particularly those with some musical knowledge. In music the first overtone is the second harmonic (for an open tube such as a flute) or the third harmonic (for a closed tube such as a clarinet). Much musical notation appears to be almost designed to confuse or trap the unwary, but it was of course developed prior to Information Theory.)

The total mean power is just the sum of the power P_n in the lines; that is

$$P = \sum_{n=0}^{\infty} P_n$$

(b) Finite Energy Signals

Since the power is zero for these signals they must be analysed in terms of their distribution of energy with frequency. This is less easy to visualise than for finite power signals. Suppose, however, we imagine a signal such as that in Figure 2.2.3 to be repeated with a fairly long period T. The power spectrum for such a repetitive signal would be similar to that in Figure 2.2.6, with lines separated by $1/T$. If we now make T progressively longer the waveform will approach the transient signal of Figure 2.2.3 and the separation between the lines will become correspondingly less. In the limit, as $T \to \infty$, the separation between the lines will approach zero and we will have a continuous spectrum. Unfortunately, in the limit, the power spectrum will have a value of zero everywhere! This is, of course, since one cannot analyse a finite energy signal in terms of power.

However, the above approach helps us to see that the energy distribution of a finite energy signal must be continuous, such as that in Figure 2.2.7. If $f(t)$ represents a voltage the units will be (mean square Volts/Hz) × time or mean square Volts/(Hz)^2. The square root of the energy spectrum is called amplitude density and has units of root mean square Volts/Hz (in this case).

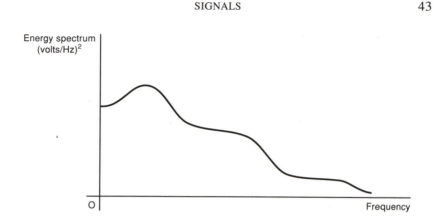

Figure 2.2.7   *Energy spectrum of a discrete signal*

Clearly, the total energy E is related to the energy spectrum $E(f)$ by

$$E = \int_0^\infty E(f)\mathrm{d}f$$

2.2.3   *Summary of Time and Frequency Domain Averages*

The results deduced above are summarised in Table 2.2.1.

Table 2.2.1

Type of signal		Time domain	Frequency domain
Finite power	continuous (random)	$\bar{f}$, $\overline{f^2}$, σ^2 etc.	continuous power spectrum
	repetitive	$\bar{f}$, $\overline{f^2}$, σ^2 etc.	discrete, lines at $1/T$, $2/T$ etc.
Finite energy		area, total energy E	continuous energy spectrum

It is important to note that the simple time averages introduced here do not give any information about the frequency content of the signals that they describe. More advanced time averages, which do not have this limitation, are discussed in section 2.3. In addition, both the time domain averages and the frequency domain averages so far considered are in no sense unique to a particular signal, since many different signals could have the same mean value or power spectrum. However, a unique relation between the time domain and frequency domain representations of signals does exist, and is introduced in section 2.4.

2.2.4   Examples for Solution

1. Figure 2.2.8 shows a repetitive triangular waveform.
   (i)   Find the area and total energy of one pulse of the waveform.
   (ii) Find the mean, mean square and variance of the repetitive waveform.

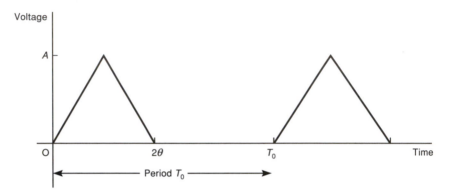

Figure 2.2.8   *Repetitive triangular waveform*

2. Figure 2.2.9 shows a pseudo-random binary sequence which repeats after a period T. Find the mean value, mean square value and variance, and discuss (and sketch) the form of its power spectrum.

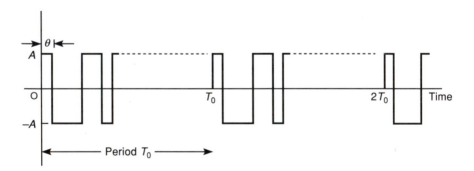

Figure 2.2.9   *Repetitive pseudo-random sequence*

2.3   Correlation and Convolution

2.3.1   Introduction

We saw above that the basic time domain averages – the mean, mean square and variance – do not provide any information about the frequency content of signals; also they do not uniquely specify a particular signal, since many different signals may have the same mean or mean square value. The processes of correlation and convolution are more complex time averages involving one or more signals and are in general dependent on the frequency content of the signals.

The general forms of the correlation function between two signals $f_1(t)$ and $f_2(t)$ (which may, as a special case, be identical) are:

finite energy signals $f(\tau) = \displaystyle\int_{-\infty}^{\infty} f_1(t) \, f_2(t + \tau) dt$

finite power signals $f(\tau) = \displaystyle\lim_{T \to \infty} \frac{1}{T} \int_{-T/2}^{T/2} f_1(t) \, f_2(t + \tau) dt$

The process of convolution is a modification of these equations in which $f_2(t)$ is time-reversed before the operation so that $f_2(-t + \tau)$ replaces $f_2(t + \tau)$.

The representation $f(t + \tau)$ denotes a time-shifted version of a function $f(t)$, τ being the amount of shift. This idea is very important and is illustrated in Figure 2.3.1 which shows a function $f(t)$ together with shifted versions of the same signal $f(t + \tau_1)$ and $f(t - \tau_2)$. The left-hand edge of $f(t + \tau_1)$ occurs earlier in time than the same edge of $f(t)$. $f(t + \tau_1)$ is said to be advanced by τ_1; similarly $f(t - \tau_2)$ is delayed by τ_2.

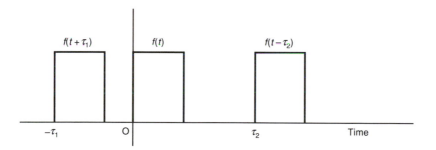

Figure 2.3.1 *Delayed and advanced pulses*

The meaning of the expression for correlation is thus that a function $f_1(t)$ is multiplied by a shifted version of a second function $f_2(t)$ and the result integrated over all time. This yields a particular value of the function $f(\tau)$ corresponding to the shift used, and $f(\tau)$ is just the set of values for all possible shifts.

We will confine our attention mainly to finite energy signals (partly because they are easier to sketch), though the equations and properties of finite power signals correspond exactly, of course.

2.3.2 Autocorrelation

As the name implies, autocorrelation is the correlation of a function with itself, so that the two functions $f_1(t)$ and $f_2(t)$ above are the same. The autocorrelation function provides a measure of the degree to which a future value of the function can be predicted, and is closely related to the energy spectrum of the signal. It is denoted by $R(\tau)$ (finite mean power) or $R_1(\tau)$ (finite energy). Sometimes $R_1(\tau)$ is denoted by $\varphi_{11}(\tau)$. For finite-energy signals

$$R_1(\tau) = \varphi_{11}(\tau) = \int_0^\infty f(t)\, f(t + \tau)\, dt$$

The correlation process is often denoted by ∇ so $R_1(\tau) = \varphi_{11}(\tau) = f_1 \nabla f_2$.

EXAMPLE 2.3.1

The autocorrelation of the rectangular pulse of Figure 2.3.2.

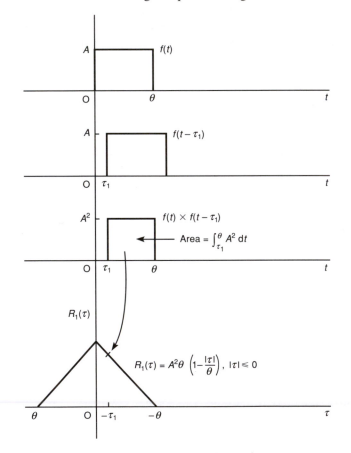

Figure 2.3.2 *Autocorrelation of a rectangular pulse*

The shift τ_1 shown produces a single point on the graph of $R_1(\tau)$. If the procedure is repeated for all τ the complete function $R_1(\tau)$ is obtained. One simply imagines moving $f(t)$ over itself to the right and left, and plots below the area obtained at each shift. Since a shift left represents delay $(-\tau)$ it is easier to plot $R_1(\tau)$ with its axes reversed, that is, $-\tau$ to the right. It is clear that $R_1(\tau)$ is always an even function (since shifting one version of $f(t)$ to the right is the same as shifting another version to the left). Also, $R_1(0)$ is the total energy (or mean power for $\bar{f^2}$ finite-power signals).

For finite-power signals it can easily be seen that $R(\tau)$ for a d.c. level is another d.c. level (implying perfect prediction), $R(\tau)$ for a sine wave is a cosine wave, and $R(\tau)$ for random noise is an impulse (zero prediction).

2.3.3 Cross-correlation

The cross-correlation function $\varphi_{12}(\tau)$ measures the similarity between two functions $f_1(t)$ and $f_2(t)$, and is given by

$$\varphi_{12}(\tau) = \int_{-\infty}^{\infty} f_1(t)f_2(t + \tau)dt = f_1 \nabla f_2$$

The order *does* matter here so $f_1 \nabla f_2$ means that f_2 is to be shifted.

EXAMPLE 2.3.2

The cross-correlation of the two rectangular pulses of Figure 2.3.3, shown on page 48.

As for autocorrelation, one function is slid to the right and left and the resulting area evaluated, $\varphi_{12}(\tau)$ being most conveniently plotted with axes reversed. It can be seen that $\varphi_{12}(\tau)$ and $\varphi_{21}(\tau)$ are mirror images in the y-axis; this is a general result, that is $\varphi_{12}(\tau) = \varphi_{21}(-\tau)$.

2.3.4 Convolution

The expression for the convolution $c_{12}(\tau)$ between two finite energy signals $f_1(t)$ and $f_2(t)$ is

$$c_{12}(\tau) = \int_{-\infty}^{\infty} f_1(t)f_2(-t + \tau)dt$$

The operation is denoted by * so $c_{12}(\tau) = f_1 * f_2$, f_2 being the function to be shifted. On first meeting the convolution function it is a little difficult to avoid the thought that it must be a pure mathematician's delight (pure mathematicians have curious tastes), since there appears to be no particular reason for choosing to reverse $f_2(t)$ before shifting it, other than it is a further possibility. However, convolution turns out to be much more important and useful than either autocorrelation or cross-correlation, and even better, its properties are relatively simple. Convolution is best thought of as a 'blurring' of one function by another.

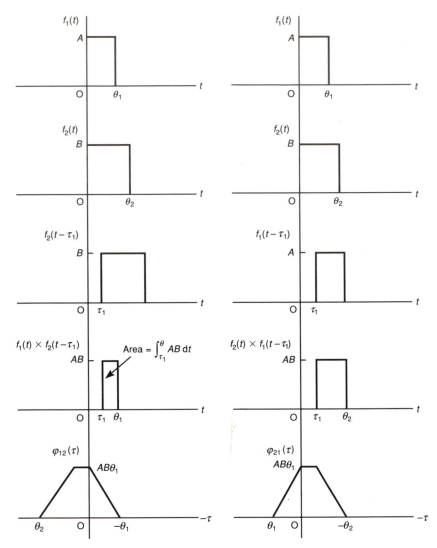

Figure 2.3.3 *Cross-correlation of two rectangular pulses*

EXAMPLE 2.3.3

The convolution of two rectangular pulses (the same two as in example 2.3.2).

The procedure is similar to correlation. except that the second function must be reversed before being shifted. The reversal has the effect that a shift of $+\tau$ now produces movement to the right (that is, delay instead of advance). However, all one has to do is slide the reversed function over the

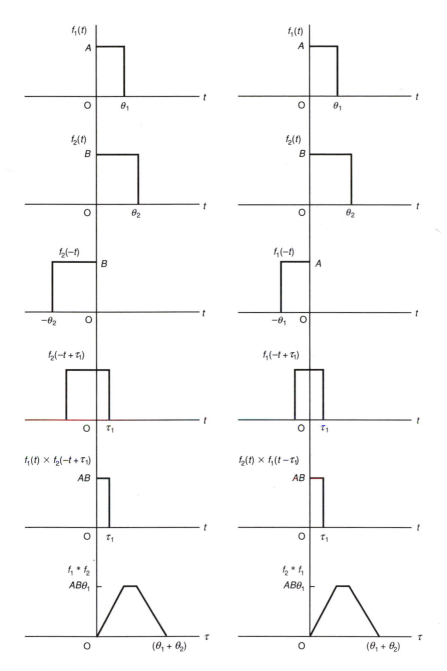

Figure 2.3.4 *Convolution of two rectangular pulses*

other and plot the area below, but now labelling the axes without reversal. One finds that, because of the time reversal, $f_1 * f_2$ is always the same as $f_2 * f_1$ (see Figure 2.3.4). The idea of convolution as 'blurring' can be seen by convolving a rectangular pulse with successively narrower pulses. This is shown in Figure 2.3.5 on page 51.

To avoid $f_1 * f_2$ becoming small as $f_2(t)$ is decreased, the area of the pulse must be kept constant, that is, $B\theta_2 = $ constant. As θ_2 approaches zero, B becomes very large, and in the limit we have $\theta_2 \rightarrow 0$, $B \rightarrow \infty$ but with $B\theta_2 = $ constant. This is known as an impulse of strength $B\theta_2$ units, being a *unit impulse* if $B\theta_2 = 1$.

We can see that if a function is convolved with a unit impulse it is *reproduced exactly at the position of the impulse*. The greater the departure from an impulse the more the function is 'blurred'. An important use of convolution with impulses is that we can construct a repetitive function by convolving one of our required pulses with a train of equally spaced impulses at the repetition period and this provides a valuable link between finite-energy and finite-power signals.

2.3.5   Applications of Correlation and Convolution

It has been necessary to introduce these functions in a fairly abstract manner at this point, since an understanding of them is required in later chapters, but it is appropriate to outline some of their applications here.

Autocorrelation is a measure of the extent to which the future value of a function can be predicted. Its main importance in information and communication is that it is related to the energy spectrum of a signal (power spectrum for finite power signals). It will be shown later that the autocorrelation function $R_1(\tau)$ and the energy spectrum $E(f)$ form a Fourier pair, that is, one transforms into the other and vice versa (the same applies to $R(\tau)$ and $P(f)$). In many communications applications this provides the easiest way (often, the only way) of finding the energy or power spectrum. For example, one cannot write an analytical formula for a binary telegraph signal, because it may be either '0' or '1' with equal probability. However, it is possible to find its autocorrelation function (see section 2.3.6, exercise 3) and hence its power spectrum.

Cross-correlation is a measure of the similarity between two functions. It is clearly of direct relevance in pattern recognition applications, where for example a read-in character or pattern may be compared with all the possible (stored) patterns to determine which it most clearly resembles. A similar application is in the detection of a signal buried in random noise, illustrated in Figure 2.3.6. Suppose the signal consists of a known shape, say a rectangular pulse, but is hidden in a large amount of noise. If we receive a sample of noise, we can test whether it contains the signal by cross-correlating the sample with our known signal shape. If the signal is present the result will

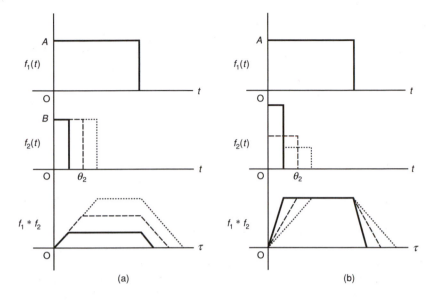

Figure 2.3.5   *Convolution as 'blurring': (a) constant height, (b) constant area*

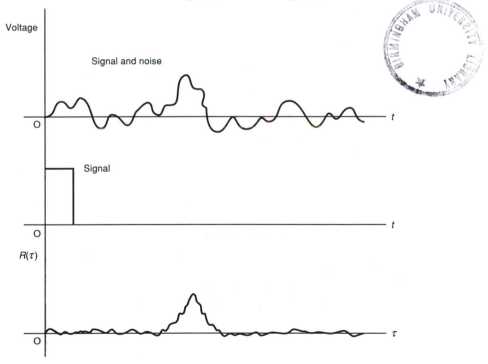

Figure 2.3.6   *Detection of a signal in noise by cross-correlation*

have a triangular peak at the position of the signal. This is strictly an example of signal detection as opposed to signal recovery.

Convolution is the blurring of one signal by another. This interpretation is borrowed from the terminology of spatial signals. In an optical system in which a transparency is projected on to a screen, some blurring of the image is always present. If the transparency consisted of a very small central pinhole the system should produce a corresponding small bright point on the screen; in practice a small area of light will be seen. This is known as the impulse response of the system, that is the output for an impulse as input, since the pinhole is the spatial equivalent of an impulse. An ideal impulse response is, of course, an impulse, but a practical one is a small roughly circular area in this case. When a transparency is used as input, the output on the screen is the two-dimensional convolution between the transparency and the impulse response of the system. Every point on the screen is blurred by the non-ideal impulse response of the system.

This is an example of one of the most fundamental properties of the convolution function; for any linear system the output is the convolution between the input and the impulse response. Another very important property is that when two functions are multiplied their Fourier Transforms are convolved, and vice versa. We will be amplifying and using both these properties later.

2.3.6   Examples for Solution

1. Answer the following, illustrating by sketches as appropriate:
   (i) Is $f_1 \nabla f_1$ necessarily even and positive?
   (ii) What is the relation between $f_1 \nabla f_2$ and $f_2 \nabla f_1$?
   (iii) What is the relation between $f_1 * f_2$ and $f_2 * f_1$?
   (iv) Is $f_1 \nabla f_1$ the same as $f_1 * f_1$?
2. Sketch the convolution of an exponential decay ($\exp(-\alpha t)$, $t > 0$) with
   (i) a rectangular pulse and (ii) a burst of sine waves. What is the physical significance of these results?
   (*Note*: the impulse response of a first-order system, such as an R–C low-pass filter, is an exponential decay.)
3. Sketch the autocorrelation function for a repetitive pseudo-random sequence (that is, a random binary telegraph signal) with levels of 0 V and 5 V, and length 8 clock pulses.
4. For the two functions $f_1(t)$ and $f_2(t)$ in Figure 2.3.7, sketch $f_1 \nabla f_1, f_2 \nabla f_2,$ $f_1 \nabla f_2, f_2 \nabla f_1, f_1 * f_1, f_2 * f_2, f_1 * f_2, f_2 * f_1$.

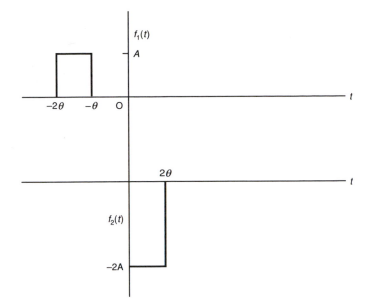

Figure 2.3.7 *Two rectangular pulses*

2.4 Fourier Theory

2.4.1 The Fourier Transform Relations

Fourier Theory uniquely relates the time and frequency domain representations of signals. It is a particularly satisfying theory, being mathematically precise and elegant, and yet having a very clear physical interpretation and many practical applications in information and communication. The topic is often introduced via Fourier Series, which apply only to finite power repetitive signals, and readers unfamiliar with the subject are referred to Appendix 3 for such an introduction.

The basic equations relating the time domain representation $f(t)$ and the frequency domain representation $F(f)$, known as the Fourier Transform relations are:

$$f(t) = \int_{-\infty}^{\infty} F(f)\exp(j\omega t)\mathrm{d}t \qquad (2.4.1)$$

$$F(f) = \int_{-\infty}^{\infty} f(t)\exp(-j\omega t)\mathrm{d}f \qquad (2.4.2)$$

The first equation is sometimes called the reverse transform and the second the forward transform, and $f(t)$ and $F(f)$ are said to constitute a transform pair. $F(f)$ is usually referred to as the spectrum of the signal $f(t)$. The equations are a very compact and pleasingly symmetrical statement of the essential idea of Fourier Theory. $f(t)$ is uniquely expressed as a sum of cisoids, of all frequencies, and $F(f)$ determines their amplitude and phase. Similarly, $F(f)$ can also be thought of as a sum of cisoids, with $f(t)$ determining amplitude and phase.

$F(f)$ is usually complex and the *amplitude spectrum* is the modulus of $F(f)$ and the *phase spectrum* its argument. It can be shown that real functions of time have an even amplitude spectrum and an odd phase spectrum. The units of the amplitude spectrum are amplitude density (the square root of energy density).

$F(f)$ can also be written as:

$$F(f) = a(f) - jb(f) = \int_{-\infty}^{\infty} f(t)(\cos\omega t - j\sin\omega t)dt$$

$$\text{so } a(f) = \int_{-\infty}^{\infty} f(t)\cos\omega t\, dt \tag{2.4.3}$$

$$\text{so } b(f) = \int_{-\infty}^{\infty} f(t)\sin\omega t\, dt \tag{2.4.4}$$

These relations are known as the sine and cosine transforms. If $f(t)$ is even then $b(f) = 0$ and similarly if $f(t)$ is odd then $a(f) = 0$, so some calculations can be simplified by using equation 2.4.3 or 2.4.4. However, we will see later that the manipulative rules of Fourier Transforms are so powerful that it is hardly ever necessary to evaluate the integral, so it is much better to think in terms of equations 2.4.1 and 2.4.2.

EXAMPLE 2.4.1

Find the spectrum of the rectangular pulse of Figure 2.4.1.

$$F(f) = \int_{-\infty}^{+\infty} f(t)\exp(-j\omega t)dt$$

$$= A \cdot \frac{-1}{j\omega} [\exp(-j\omega t)]_{-\theta/2}^{\theta/2}$$

$$= \frac{2A}{\omega} \sin\omega\theta/2$$

$$= A\theta \operatorname{sinc}(f\,\theta) \quad \text{where } \operatorname{sinc}(f\theta) \text{ is defined as } \sin(\pi f\theta)/(\pi f\theta)$$

The sinc function is very important in Fourier Theory and occurs whenever a rectangular shaped signal is involved. It is plotted in Figure A3.6 in Appendix 3. $\operatorname{sinc}(x)$ has a maximum value of unity for $x = 0$ and is zero for

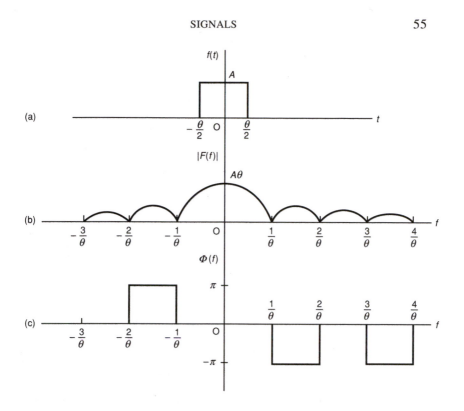

Figure 2.4.1 *A rectangular signal (a) and its spectrum (b and c)*

all other integral values, so it is far easier to plot than, say, a sine wave. It is the author's favourite (mathematical) function. The only unfortunate thing about it is that most 'intelligent' word processors kindly replace 'sinc' by 'sin c' whenever it occurs!

This is a very basic result and should be memorised. $F(f)$ is real in this case, so it could be plotted on a single diagram. However, the amplitude and phase spectra are shown separately in parts (b) and (c) of Figure 2.4.1.

EXAMPLE 2.4.2

Find the signal corresponding to the rectangular spectrum of Figure 2.4.2(a).

$$f(t) = \int_{-\infty}^{+\infty} F(f)\exp(j\omega t)df$$

$$= \int_{-f_c/2}^{f_c/2} B\exp(j\omega t)df$$

Proceeding exactly as in example 2.4.1 we obtain $f(t) = Bf_c\text{sinc}(f_c t)$.

These examples illustrate the basic symmetry of the Fourier Transform

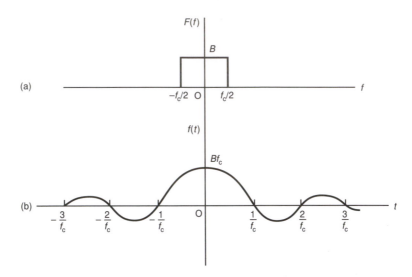

Figure 2.4.2 *A rectangular spectrum (a) and its signal (b)*

relations; a function $f(t)$ can be converted to its spectrum $F(f)$ by the forward transform, the reverse transform being used for the reverse direction. (Logically one requires only a single transform, but it requires three applications of the forward transform to convert $F(f)$ back to $f(t)$, so the reverse transform is convenient.) A Fourier pair, that is $f(t)$ and its corresponding $F(f)$ are often denoted by

$$f(t) \Leftrightarrow F(f)$$

Note that the zero-frequency value of $F(f)$ is equal to the pulse area. This is a useful general result, i.e.

$$\int_{-\infty}^{+\infty} f(t)\,dt = F(0) \quad \text{and conversely} \quad \int_{-\infty}^{+\infty} F(f)\,df = f(0) \quad (2.4.5)$$

It can be seen from these examples that if a function is extended in one domain it is narrow in the other, and vice versa. This is known as the principle of reciprocal spreading. It has a direct application in optics, where the spectrum of light from an incandescent source is broad because the light is emitted in short bursts, whereas that from a laser is narrow because the wavetrains are very long. There is an approximate general relation between the 'width' of a function Δt in the time domain and its corresponding 'width' Δf in the frequency domain:

$$\Delta f \times \Delta t \approx 1$$

(The 'width' really requires definition, and is the range from the origin to the first zero crossing in the above examples.)

One important class of functions, which has the same form in each domain, is the Gaussian $\exp(-at^2)$ or $\exp(-Af^2)$. This again has an application in optics; the intensity distribution across a laser beam is Gaussian, and transforms performed on it (by lenses etc., see later) do not change the form (though they may change the scale).

We can also deduce the spectrum of an impulse, by letting A tend to infinity and θ tend to zero, but with the area $A\theta$ constant. $F(f)$ is constant with value equal to the strength, that is, unity for a unit impulse. It is therefore a useful function for measuring the frequency response of a system, because its application is equivalent to applying all frequencies simultaneously with equal amplitude and zero phase.

EXAMPLE 2.4.3

Find the spectrum of the exponential decay of Figure 2.4.3(a).

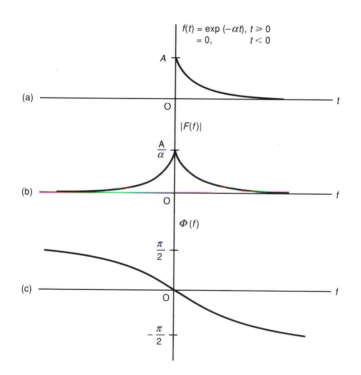

Figure 2.4.3 *An exponential decay (a) and its spectrum (b and c)*

This function is neither odd nor even, so we have no choice but to use the full transform here.

$$F(f) = \int_0^\infty A\exp(-\alpha t)\exp(-j\omega t)dt$$

$$= \frac{A}{\alpha + j\omega}$$

Thus

$$|F(f)| = \frac{A}{(\alpha^2 + \omega^2)^{\frac{1}{2}}}$$

$$\Phi(f) = -\tan^{-1}\frac{\omega}{\alpha}$$

These are plotted in Figure 2.4.3(b) and (c).

We can deduce the spectrum of the unit step function from this example, by letting α tend to zero, so $F(f) \to 1/j\omega$. This is not strictly correct; the Fourier Transform expressions 2.4.1 and 2.4.2 are valid only when the integrals converge, and a requirement for this is that $\int_{-\infty}^{+\infty} f(t)dt$ is finite. A unit step is not, of course, a finite energy function, and its area is similarly infinite. It can be shown that its Fourier Transform is actually $1/j\omega + \frac{1}{2}\delta(0)$, that is, an impulse of strength one-half at zero frequency is present in addition to the $1/j\omega$ term. In fact, $1/j\omega$ is the spectrum of the signum function, shown in Figure 2.4.4.

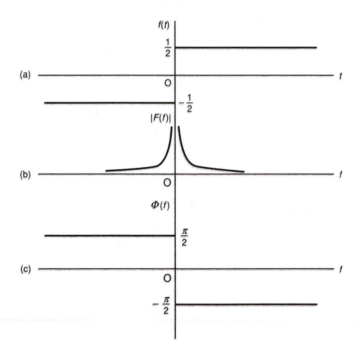

Figure 2.4.4   *The signum function (a) and its spectrum (b and c)*

2.4.2   Operator Form of Fourier Transforms

The process of performing a forward transform on a signal $f(t)$ to obtain its spectrum $F(f)$ may be viewed as a mathematical operation and denoted by an operator T, defined as follows:

$$T\{f(t)\} = \int_{-\infty}^{\infty} f(t)\exp(-j\omega t)\mathrm{d}t$$

so that $F(t) = T\{f(t)\}$. The operation consists of multiplying the operand by $\exp(-j\omega t)$ and integrating over all values with respect to the operand. Performing $T\{f(t)\}$ once we obtain $F(f)$, of course, but if we now perform the operation again, the result is

$$T^2\{f(t)\} = T\{F(f)\} = \int_{-\infty}^{\infty} F(f)\exp(-j\omega t)\mathrm{d}f = f(-t)$$

which is the mirror image of $f(t)$ in the vertical axis. Similarly, a third operation produces $F(-f)$ and a fourth yields the original signal $f(t)$. It is convenient to introduce a 'unity' operator I (which does nothing) and a reversing operator $-I$ (which produces a mirror image of the input). The above results can then be expressed as $T^2 = -I$ and $T^4 = I$.

The operator T actually refers to the reverse transform (equation 2.4.1). A second operator T_2 could be defined for the forward transform in a similar manner, that is $T_2\{F(f)\} = \int_{-\infty}^{\infty} F(f)\exp(j\omega t)\mathrm{d}f$. However, it can be seen that this operation (multiplying by $\exp(j\omega t)$ and integrating with respect to the operand) is just the complex conjugate of T, so the forward transform operator T_2 is denoted T^* and is given by

$$T^*\{f(t)\} = \int_{-\infty}^{\infty} f(t)\exp(j\omega t)\mathrm{d}t$$

The results of repeated operations of T and T^* are shown in Table 2.4.1, in which movement to the right indicates the effect of T and movement to the left indicates T^*. Figure 2.4.5 illustrates the effects geometrically.

Table 2.4.1

$f(t) \Leftrightarrow F(f) \Leftrightarrow f(-t) \Leftrightarrow F(-f) \Leftrightarrow f(t) \Leftrightarrow F(f)$ etc.
$f^*(t) \Leftrightarrow F^*(-f) \Leftrightarrow f^*(-t) \Leftrightarrow F^*(f) \Leftrightarrow f^*(t) \Leftrightarrow F^*(-f)$ etc.

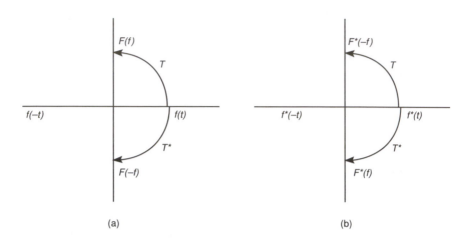

(a)   (b)

Figure 2.4.5 *Effect of repeated operations of T and T* on (a) f(t) and
(b) f*(t)*

2.4.3 Fourier Transforms of Spatial Signals

Although we have so far considered only temporal signals, Fourier Transforms occur between corresponding variables in several other fields, such as
fluid flow and optical systems, and the basic expressions 2.4.1 and 2.4.2 can
be generalised in terms of variables x (replacing t) and X (replacing f) producing the equations below:

$$f(x) = \int_{-\infty}^{\infty} F(X)\exp(j2\pi xX)dX$$
$$F(X) = \int_{-\infty}^{\infty} f(x)\exp(-j2\pi xX)dx$$

The interpretation is of course exactly the same, $f(x)$ being expressed as a
sum of cisoids $\exp(j2\pi xX)$ with $F(X)$ determining their amplitude and phase.

We will consider here only optical systems, since the communication of
spatial data is of course of great importance. An interesting analogy exists
between electronic and optical systems. Voltage is analogous to transmitted
light intensity and time analogous to distance, so the illuminated slit in Figure 2.4.6(a) has the transmission function shown in Figure 2.4.6(b) and is
equivalent to the rectangular pulse of Figure 2.4.6(c).

Figure 2.4.7 shows an optical system, in which coherent light from a
laser illuminates a transparency, producing an image on the screen.

The light transmitted by the transparency is in the form of a spatial distribution of light intensity $f(x, y)$; in each dimension a distance x (or y) is
analogous to the time variable of a function $f(t)$. If such a transparency is
placed in the front focal plane of a lens it can be shown that the (two-
dimensional) Fourier Transform appears in the back focal plane. What actu-

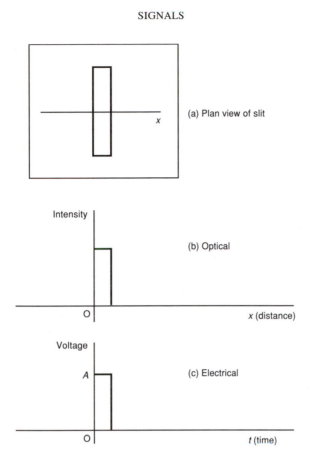

Figure 2.4.6 *Optical and electrical analogies*

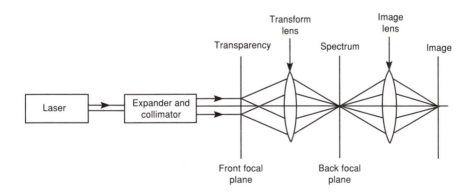

Figure 2.4.7 *Apparatus for producing an optical Fourier Transform*

ally happens is that the transparency diffracts the light (producing phase differences dependent on the angle of diffraction) and the transform lens collects it again (which is equivalent to integration).

The phase differences are such that an exact Fourier Transform is produced if appropriate variables are chosen. Choosing x and y for distances in the plane of the transparency and $X = u/\lambda L$, $Y = v/\lambda L$ (where u and v are distances in the focal plane of the lens, λ is the wavelength of the light and L is the focal length of the lens) the two-dimensional Fourier Transform equation is

$$F(X, Y) = \int_{-\infty}^{\infty} \int_{-\infty}^{\infty} f(xy)\exp(-j2\pi(xX + yY))dx\, dy$$

There is only one equation in this case but successive transforms can be carried out by successive lenses in the optical system.

The spatial equivalent of frequency (X or Y) is called *spatial frequency* and has dimensions of lines/metre. If, for example, the transparency was a diffraction grating (a one-dimensional signal) the diffracted orders would appear as bright spots in the back focal plane. A diffraction grating is the optical equivalent of a sequence of rectangular pulses, so the intensity of the spots follows a sinc function as shown in Figure 2.4.8. The spatial frequency of the grating is so many lines/metre and the distance of a given spot from the axis is directly proportional to its spatial frequency. For a two-dimensional transparency, the light distribution in the back focal plane is a two-dimensional display of the spatial frequencies present in the transparency.

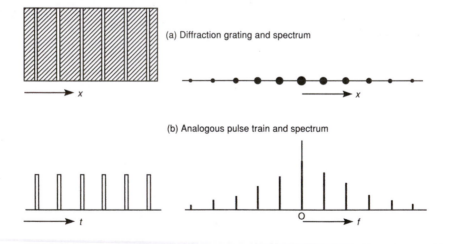

(a) Diffraction grating and spectrum

(b) Analogous pulse train and spectrum

Figure 2.4.8   *Diffraction grating and its spectrum*

An interesting consequence of this electronic/optical analogy is that the transform of the transparency is directly physically observable (unlike the electronic system) so that one can inspect object and transform together and operate on the spectrum before producing the image. This entails blocking out parts of the spectrum (spatial filtering). For example, a low-pass filter is simply a card with a central hole placed in the back focal plane.

The fact that a Fourier Transform can be performed optically has found a number of applications, such as in removing unwanted patterns from photographs (for example, from fingerprints taken from fabric), checking metallic gratings for use in image tubes, deblurring photographs, in character recognition and in data analysis. There are some experimental difficulties in that a laser must be used to produce the transform, and the data to be processed must be in a suitable form (such as a transparency). However, the technique is likely to grow in importance as input mechanisms are improved, and has inherent advantages over electronic systems in that a two-dimensional transform of very high resolution can be produced instantaneously (that is, at the speed of light).

2.4.4 Properties of Fourier Transforms

Fourier Transforms have a number of important and powerful properties, which greatly simplify the process of deducing transforms for most functions. The list given here is by no means exhaustive; only those essential to information and communication are included. If we omit the obvious rules for addition of two functions (when the Transforms are added) and multiplication by a constant (when the Transform is multiplied by the same constant) there are actually only three important properties. It would not be appropriate to give formal proofs here but such proofs can be found in Stuart (1966).

(i) *Time shift*

$$\text{If } f(t) \Leftrightarrow F(f) \text{ then } f(t-\tau) \Leftrightarrow F(f) \exp(-j\omega\tau)$$

This can be proved formally from relationship 2.4.1 and can be verified, for example, by calculating the transform of a pulse such as in example 2.4.1 with its origin at the edge. The relation states that a delay or advance τ simply adds a phase term $\exp(j\omega\tau)$ to the spectrum.

(ii) *Differentiation*

$$\text{If } f(t) \Leftrightarrow F(f) \text{ then } \frac{df(t)}{dt} \Leftrightarrow j\omega F(f)$$

This can also be proved directly from expression 2.4.1. It states that the operation of differentiating a function is equivalent to multiplying by $j\omega$ in the frequency domain.

(iii) *Convolution*

Assuming again that $f_1(t) \Leftrightarrow F_1(f)$ and $f_2(t) \Leftrightarrow F_2(f)$:

(a) $F_1(f) \times F_2(f) \Leftrightarrow f_1(t) * f_2(t)$
(b) $f_1(t) \times f_2(t) \Leftrightarrow F_1(f) * F_2(f)$

These rules mean that if a signal is equal to the product of two other signals, then its spectrum is the convolution between the corresponding spectra, and similarly for the convolution of two signals. The operations of multiplication and convolution are complementary in Fourier Transforms. (These can again be proved from expression 2.4.1, but the proof is somewhat lengthy and tedious.)

The following examples illustrate the various rules.

EXAMPLE 2.4.4

Find the spectrum of the pair of pulses of Figure 2.4.9(a).

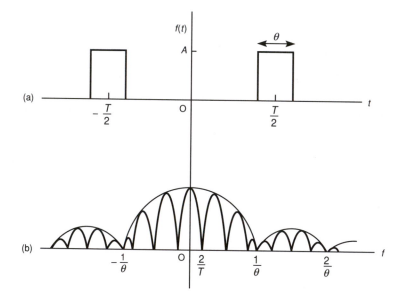

Figure 2.4.9 *Two pulses (a) and their amplitude spectra (b)*

Using the rules for time shift and addition

$$F(f) = A\theta \; \text{sinc} \; f\theta \; \exp(-j\omega T/2) + a\theta \; \text{sinc} \; f\theta \; \exp(+j\omega T/2)$$
$$= A\theta \; \text{sinc} \; f\theta.2\cos \pi fT$$

The spectrum is similar to that for a single pulse but with 'beats' due to the second pulse. In fact it is identical to the interference seen in optics between two slits illuminated by coherent light.

EXAMPLE 2.4.5

The spectrum of the triangular pulse of Figure 2.4.10(a).

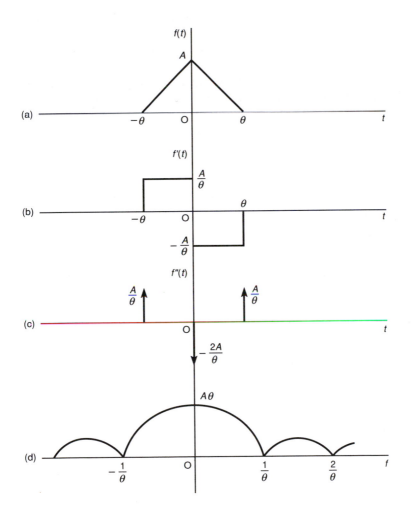

Figure 2.4.10 (a) *Triangular pulse f(t), (b) differentiated once,*
(c) differentiated twice, (d) spectrum

There are at least four different ways of doing this example. The most obvious is to let $f(t) = A\left(1 - \dfrac{|t|}{\theta}\right)$, $|t| < 0$ for the triangle and use the standard formula; that is

$$F(f) = \int_{-\theta}^{\theta} A\left(1 - \frac{|t|}{\theta}\right)\exp(-j\omega t)dt$$

$$\theta$$

This is without doubt the world's worst way of solving this problem! Although the integral is not very difficult, the working takes about a page of A4 and the probability of error is rather high (about 99.9 per cent for most students). It should *never* be done this way.

A much better method is to differentiate $f(t)$, producing rectangular pulses whose transforms are known, as shown in Figure 2.4.10(b).

$$j\omega F(f) = \frac{A}{\theta} \cdot \theta \text{ sinc } f\theta \cdot \exp(j\omega\theta/2) - \frac{A}{\theta} \cdot \theta \cdot \text{ sinc } f\theta \cdot \exp(-j\omega\theta)/2)$$

$$= A \text{ sinc } f\theta(2j\sin \omega\theta/2)$$

so $F(f) = A\theta \text{ sinc}^2 f\theta$

Alternatively, $f(t)$ can be differentiated twice, when only impulses remain, as in Figure 2.4.10(c). Using the fact that a unit impulse at the origin has a transform of unity:

$$(j\omega)^2 F(f) = \frac{A}{\theta} (\exp(j\omega\theta) - 2 + \exp(-j\omega\theta))$$

$$= \frac{2A}{\theta} (\cos \omega\theta - 1)$$

$$= \frac{-4A}{\theta} \sin^2 \frac{\omega\theta}{2}$$

so $\quad F(f) = A\theta \text{ sinc}^2 f\theta \quad$ as before

However, the best method is to use convolution. The triangle can be seen to be the convolution between two rectangular pulses, each of width θ (and therefore having a spectrum $A\theta \text{ sinc } f\theta$). In order that the height of the triangle be A, the pulses must have heights of $\sqrt{(A/\theta)}$. We thus have

$$F(f) = \sqrt{\frac{A}{\theta}} \cdot \theta \text{ sinc } f\theta. \sqrt{\frac{A}{\theta}} \cdot \theta \text{ sinc } f\theta = A\theta \text{ sinc}^2 f\theta$$

Actually it is even easier than this. We do not have to bother about the height of the pulses since we know from equation 2.4.5 that the area of the triangle must be equal to the zero-frequency component of its spectrum. In one line, therefore

$$F(f) = (\text{area}) \text{ sinc } f\theta \cdot \text{ sinc } f\theta = A\theta \text{ sinc}^2 f\theta$$

Table 2.4.2 summarises the important properties of Fourier Transforms and includes the spectra of some important signals in information and communication.

Table 2.4.2 *Table of Fourier Transforms*

	Signal	*Spectrum*
Basic relations	$f(t)$	$F(f)$
	$f(-t)$	$F(-f)$
	$f^*(t)$	$F^*(-f)$
	$f^*(-t)$	$F^*(f)$
Properties	$a\,f(t)$	$a\,F(f)$
	$f_1(t) + f_2(t)$	$F_1(f) + F_2(f)$
	$f(t - \tau)$	$F(f)\exp(-j\omega\tau)$
	$f'(t)$	$j\omega F(f)$
	$f_1 \times f_2$	$F_1 * F_2$
	$f_1 * f_2$	$F_1 \times F_2$
Important signals	$\delta(t), 1$	$1, \delta(f)$
	$\mathrm{rect}(t), \mathrm{sinc}(t)$	$\mathrm{sinc}(f), \mathrm{rect}(f)$
	$\exp(-\pi t^2)$	$\exp(-\pi f^2)$
	$\cos(\omega_0 t)$	$\frac{1}{2}\delta(f - f_0) + \frac{1}{2}\delta(f + f_0)$
	$\sin(\omega_0 t)$	$-\frac{1}{2}j\,\delta(f - f_0) + \frac{1}{2}j\delta(f + f_0)$
	$\exp(j\omega_0 t)$	$\delta(f - f_0)$
	$\exp(-\alpha t)$	$1/(\alpha + j\omega)$
	rectangular pulse A, θ	$A\theta\,\mathrm{sinc}(f\theta)$
	triangular pulse A, 2θ	$A\theta\,\mathrm{sinc}^2(f\theta)$
	train of impulses $\delta(t - nT_0)$	train of impulses $1/T \times \delta(f - nf_0)$

2.4.5 Repetitive Signals

Since a repetitive signal of period T can be thought of as the convolution between a single pulse and a set of impulses spaced by T, we can find the Fourier Transform of a repetitive signal if we know that of a set of impulses. It is shown in Appendix 3 that the Fourier Series spectrum of a set of unit impulses is a set of lines of height $1/T$ spaced at $1/T$. It can be shown that the Fourier Transform equivalent is simply a set of impulses of strength $1/T$ spaced at $1/T$. In fact, if we have the Fourier Series representation of a repetitive signal (always a set of lines spaced at $1/T$) then the Fourier Transform representation is the same, but with impulses replacing the lines. The spectrum can be found from first principles by adding successive impulses with appropriate delays, but the proof is very tedious. We thus have the important result shown in Figure 2.4.11.

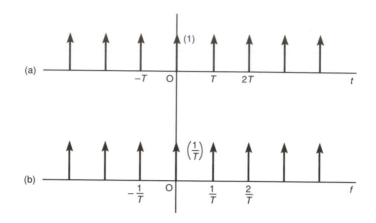

Figure 2.4.11 *Sequence of impulses (a) and its spectrum (b)*

EXAMPLE 2.4.6

Find the spectrum of the repetitive pulse train of Figure 2.4.12(a).

The repetitive train is the convolution between a set of impulses and a single rectangular pulse. The spectrum is therefore the product between the spectrum of the impulses (impulses of strength $1/T$ and spacing $1/T$) and that of the single pulse (a sinc function).

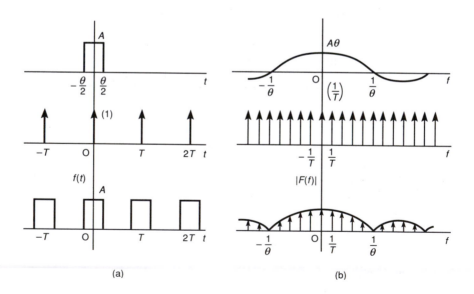

Figure 2.4.12 *Repetitive pulse train (a) and its spectrum (b)*

EXAMPLE 2.4.7

Find the spectrum of the limited train of sine waves of Figure 2.4.13(a).

The spectrum of a continuous cosine wave consists of two impulses of strength one-half, since $\cos \omega_0 t = \dfrac{1}{2}(\exp(j\omega_0 t) + \exp(-j\omega_0 t))$ and the Fourier Transform is simply a plot of cisoids. A limited train of sine waves of length θ can be obtained by multiplying a continuous train by a rectangular pulse of length θ. The spectrum of the limited train is therefore the convolution between the spectrum of the continuous train (two impulses) and that of the rectangular pulse (a sinc function).

$$F(f) = \frac{A\theta}{2}\left[\ \text{sinc}\left(f + \frac{1}{T}\right) + \text{sinc}\left(f - \frac{1}{T}\right)\right]$$

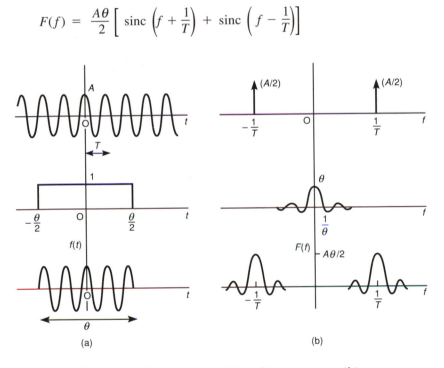

Figure 2.4.13 *Burst of sine waves (a) and its spectrum (b)*

2.4.6 The Discrete Fourier Transform and the Fast Fourier Transform

In order to compute the Fourier Transform of a function one must obviously convert the signal into a (finite) number of samples. Assume that the signal has duration T and that N samples are taken at equal intervals T/N, so the discrete function f_i replacing $f(t)$ has values at $0, T/N, 2T/N \ldots \ldots$ $(N - 1)T/N$. Since the duration is T the frequency values will be at intervals of $1/T$ so the discrete function F_n replacing $F(f)$ has values at

0, $1/T$, $2/T$. . . . . . $(N - 1)/T$. Replacing the integrals of the Fourier Transform relations 2.4.1 and 2.4.2 by summations, putting $F_n = F(f)/T$ and replacing df by $1/T$ and dt by T/N we obtain the *Discrete Fourier Transform* (DFT) relations:

$$f_i = \sum_{n=0}^{N-1} F_n \exp(j2\pi in/N)$$

$$F_n = \frac{1}{N} \sum_{i=0}^{N-1} f_i \exp(-j2\pi in/N)$$

The summation is over the range 0 to $N - 1$, rather than $-N/2$ to $+N/2$, for both i and n to simplify computation. Because we are dealing with discrete values both f_i and F_n are effectively periodic, so they repeat themselves outside the ranges 0 to T (for f_i) and 0 to N/T (for F_n); this is why $F(f)$ was replaced by F_n/T, in the same way as $F(f) = c_n T$ when deriving the Fourier Transform relations from Fourier Series, as in Appendix 3.

The sampling rate $R = N/T$ per second and it can be shown (see section 2.5) that the highest frequency at which the computation is valid is $R/2$. The modulus of F_n for values greater than $R/2$ ($N/2T$) is the mirror image of the values in the lower range. Figure 2.4.14 shows a sampled rectangular pulse

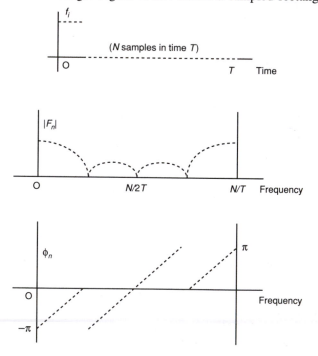

Figure 2.4.14   *Discrete Fourier Transform of a rectangular pulse*

and the corresponding modulus and phase of F_n.

The *Fast Fourier Transform* (FFT) is an efficient computer algorithm for the evaluation of the Discrete Fourier Transform, and may be as much as two orders of magnitude faster than a direct implementation of the summations. It was invented in the 1960s and employs a technique known as base 2 decimation in frequency with reverse binary bit re-ordering at the beginning. It provides a transformation between N pairs of real and imaginary numbers. The data are entered into an array (for a real function the imaginary parts are all zero, of course) and the transformation replaces the elements by the corresponding transformed real and imaginary elements. Both forward and reverse transforms are possible.

It is important to realise that the DFT is simply a discrete version of the Fourier Transform relations of 2.4.1 and 2.4.2, and the FFT a computer implementation of the DFT; they are included here for completeness. If one needs to know the spectrum of a waveform in a communication system, one uses 2.4.1 and 2.4.2 of course, irrespective of whether the waveform is continuous or sampled.

2.4.7 Energy/Power Spectra and Autocorrelation

The ideas of Energy Spectrum, Power Spectrum and Autocorrelation were introduced in section 2.3. We can now relate them to the Fourier Theory expressions.

Finite Energy Signals

The energy E in a finite energy signal is given by

$$E = \int_{-\infty}^{\infty} f^2(t)\mathrm{d}t = \int_{-\infty}^{\infty} E(f)\mathrm{d}f \quad \text{(taking } E(f) \text{ two-sided)}$$

Therefore

$$
\begin{aligned}
E &= \int_{-\infty}^{\infty} f(t)[\int_{-\infty}^{\infty} F(f)\exp(j\omega t)\mathrm{d}f]\mathrm{d}t \\
&= \int_{-\infty}^{\infty} F(f)[\int_{-\infty}^{\infty} f(t)\exp(j\omega t)\mathrm{d}t]\mathrm{d}f \\
&= \int_{-\infty}^{\infty} F(f)\, F^*(f)\mathrm{d}f \qquad (F^*(f) \text{ is the complex conjugate of } F(f)) \\
&= \int_{-\infty}^{\infty} |F(f)|^2\, \mathrm{d}f
\end{aligned}
$$

However $E = \int_{-\infty}^{\infty} E(f)\mathrm{d}f$ as above so it can be seen by equating the integrands that $E(f)$ is just the squared modulus of the Fourier Transform (the phase information in $F(f)$ is therefore lost). That is

$$E(f) = |F(f)|^2 \tag{2.4.6}$$

Finite Power Signals

Similarly, for a finite power signal the signal power P given by

$$P = \lim_{T \to \infty} \frac{1}{T} \int_{-T/2}^{T/2} f^2(t)\mathrm{d}t = \int_{-\infty}^{\infty} P(f)\mathrm{d}f$$

For a repetitive signal we have $P = \sum_{n=-\infty}^{n=\infty} |c_n|^2$, and proceeding as above we

obtain

$$P_n = |c_n|^2 \tag{2.4.7}$$

so the power in the nth line is $|c_n|^2$. Continuous finite power signals have a continuous power spectrum but cannot be expressed in terms of a complex Fourier spectrum, since their phase spectrum is random (uniform).

Autocorrelation

We can find an important expression for the autocorrelation function of a finite energy signal in a similar manner.

$R_1(\tau) = \int_{-\infty}^{\infty} f(t)f(t + \tau)\mathrm{d}t$ and replacing $f(t + \tau)$ by $\int_{-\infty}^{\infty} F(f)\exp(j\omega\tau)\mathrm{d}f$ and changing the order of integration as above, we find $R_1(\tau) = \int_{-\infty}^{\infty} E(f)\exp(j\omega\tau)\mathrm{d}f$. This is exactly the Fourier Transform relation between $R_1(\tau)$ and $E(f)$, so we find that $R_1(\tau)$ and $E(f)$ are a *Fourier pair*.

Similarly, for finite power signals $R(\tau)$ and $P(f)$ are a *Fourier pair*.

Some examples of power spectra and autocorrelation functions for finite power signals are shown in Figure 2.4.15.

EXAMPLE 2.4.8

Verify the above relations for the rectangular pulse of Figure 2.4.16.

We have $F(f) = A\theta \operatorname{sinc}(f\theta)$ as before, so $E(f) = A^2\theta^2 \operatorname{sinc}^2(f\theta)$. Also $R_1(\tau) = A^2\theta(1 - |\tau|/\theta)$, $|\tau| \leq 0$. Now $E(f)$ and $R_1(\tau)$ are a Fourier pair and since $R_1(\tau)$ is triangular we can use the result derived above, so $E(f) = (\text{area}) \operatorname{sinc}^2(f\theta) = (A\theta)^2\operatorname{sinc}^2(f\theta)$ as above. Unfortunately we cannot easily evaluate the Fourier Transform of $E(f)$, so we cannot formally show that it is equal to $R_1(\tau)$.

Finally, the total energy, $E = \int_{-\infty}^{\infty} f^2(t)\mathrm{d}t = A^2\theta$, as an integral over time. Alternatively, as an integral over frequency
$E = \int_{-\infty}^{\infty} E(f)\mathrm{d}f = \int_{-\infty}^{\infty} (A\theta)^2\operatorname{sinc}^2(f\theta)\mathrm{d}f \to A^2\theta$ (since the integral of $\operatorname{sinc}^2 x$ over all values is unity).

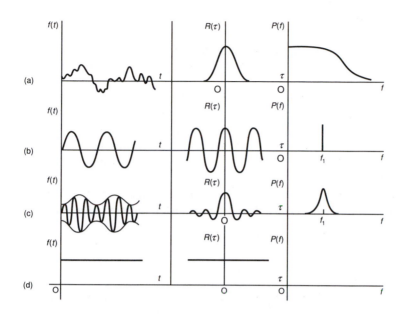

Figure 2.4.15 *Power spectrum and autocorrelation for various signals: (a) noise, (b) sine wave, (c) noisy sine wave, (d) d.c. level*

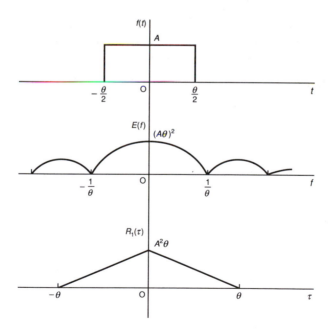

Figure 2.4.16 *Power spectrum and autocorrelation function of a rectangular pulse*

2.4.8   Linear Systems

Fourier Theory provides a valuable insight into the behaviour of linear systems, and we will conclude this section by giving a brief summary of the important relations.

A linear system obeys the *principle of superposition*, which means that if an input i_1 produces an output o_1 and an input i_2 produces an output o_2 then inputs (i_1 and i_2) together produce an output ($o_1 + o_2$). Essentially, no frequencies are present at the output that were not present at the input.

Consider the system of Figure 2.4.17 in which an input $f(t)$ produces an output $g(t)$.

In the frequency domain we have an input spectrum $F(f)$ and an output spectrum $G(f)$ related by the system transfer function $H(f)$. That is

$$G(f) = H(f) \times F(f)$$

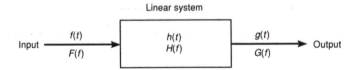

Figure 2.4.17   *Fourier relations in a linear system*

$H(f)$ is complex; its modulus is the amplitude response of the system, and its argument the phase response, so $H(f) = A(f) \exp(j\phi f))$. The time-equivalent of $H(f)$ has been labelled $h(t)$. To see its significance, let $f(t)$ be a unit impulse. We have $F(f) = 1$ so $G(f) = H(f)$. The output time response is $g(t) = h(t)$; in other words $h(t)$ is the impulse response of the system and $h(t)$ and $H(f)$ are a Fourier pair. The impulse response of a system is a particularly important parameter. It is more fundamental that the widely used step response (preferred by engineers because it is more easily realised in practice), being the Fourier Transform of the transfer function.

Since $G(f)$ is the product of $H(f)$ and $F(f)$ we can immediately deduce that

$$g(t) = h(t) * f(t)$$

that is, the output is the convolution between input and impulse response.

The impulse response has the significance of representing the system's 'memory' of past inputs. A 'broad' impulse response means that the output at a given instant is determined by previous inputs stretching backwards in time; an ideal system has an impulse response that is an impulse and the output depends only on the present input.

We therefore have two methods for finding the response of a system; we can use a convolution in the time domain, or we can use a product in the frequency domain and then carry out a Fourier Transform if we want the time response.

EXAMPLE 2.4.9

Step response of a low-pass filter.

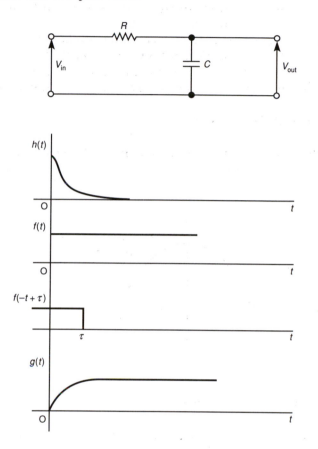

Figure 2.4.18 *Step response of a low-pass filter*

For the filter of Figure 2.4.18 $H(f) = \dfrac{V_{out}}{V_{in}} = \dfrac{\dfrac{1}{j\omega C}}{R + \dfrac{1}{j\omega C}} = \dfrac{\alpha}{\alpha + j\omega}$

(where $\alpha = 1/RC$)

The impulse response $h(t)$ is the Transform of $H(f)$ so $h(t) = \exp(-\alpha t)$ (since we know the Transform of the exponential decay).

Since $g(t) = h(t) * f(t)$ we can see geometrically that $g(t) = 1 - \exp(-\alpha t)$. Alternatively, in the frequency domain

$$G(f) = H(f) \times F(f) = \frac{1}{j\omega}\frac{\alpha}{\alpha + j\omega} = \frac{1}{j\omega} - \frac{1}{\alpha + j\omega}$$

Taking the Fourier Transform of $G(f)$, $g(t) = 1 - \exp(-\alpha t)$ for $t > 0$, taking the Transform of the unit step to be $1/j\omega$.

The same relations apply to optical systems discussed in section 2.4.3. The output $g(t)$ is the convolution between the input $f(t)$ and the impulse response $h(t)$, and similarly in the frequency domain the output can be obtained by multiplying the spectrum of the input by the transfer function.

2.4.9 Examples for Solution

1. Find the spectrum of the 'half cosine' pulse of Figure 2.4.19 and sketch its amplitude spectrum.

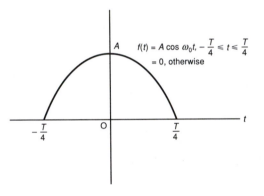

Figure 2.4.19 *Half-cosine pulse*

2. Find the spectrum of the double pulse of Figure 2.4.20. Sketch the amplitude spectrum for $T = 2\tau$ and $T = \tau/2$.

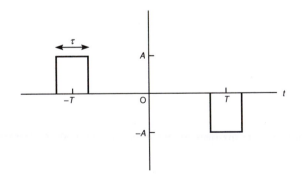

Figure 2.4.20 *Double pulse*

3. (i) Find the spectrum of the single pulse of Figure 2.4.21(a). Hence, deduce and sketch the modulus of the spectrum of the single pulse of Figure 2.4.21(b).

   (ii) The waveform of Figure 2.4.21(b) is repeated continuously with period T. Find the ratio of the amplitudes of the first six harmonics.

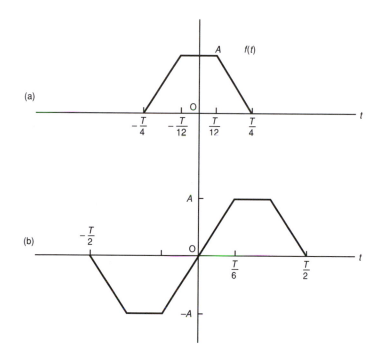

Figure 2.4.21   *Trapezoidal pulse (a) and waveform (b)*

4. A simple electronic organ produces a repetitive triangular pulse waveform of period T as shown in Figure 2.4.22. At a given frequency the basewidth 2θ of the triangle can be changed over the range 0 to T to

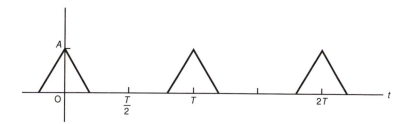

Figure 2.4.22   *Triangular waveform*

simulate the sounds of chosen musical instruments. Find a general expression for the amplitude spectrum of the waveform and sketch it for $\theta = T/4$. Estimate the values of θ, as fractions of T, to make reasonable simulations of the three instruments, whose relative harmonic content should be assumed to be as shown below.

Harmonic	Clarinet	Trumpet	Flute
1	10	10	10
2	0.1	5	1.0
3	1.1	1.1	0.4
4	0.1	0.1	0.6
5	0.4	0.4	0.1
6	0.1	0.6	0.2
7	0.2	0.2	0.1
8	0.1	0.1	0.1

5. A function of time $f(t)$ is shown in Figure 2.4.23. It consists of a series of equally spaced triangles, all of the same base-width $2a/3$, with heights determined by a larger triangle. Express $f(t)$ in terms of products and convolutions between simpler functions, and hence sketch the modulus of the Fourier Transform of $f(t)$ versus frequency.

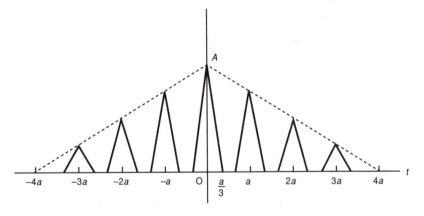

Figure 2.4.23 *The triangle of triangles*

6. Find the finite autocorrelation function $R_1(\tau)$ for the function $f(t)$ defined by

$$f(t) = A \exp(-\alpha t), \quad (t \geq 0)$$
$$= 0 \qquad\qquad (t < 0)$$

and hence find the energy spectrum $E(f)$ and the total energy E.
Find the Fourier transform $F(f)$ of $f(t)$ and show that $|F(f)|^2 = E(f)$ and that $\int_{-\infty}^{\infty} f^2(t)dt = E$. Sketch the graphs of $E(f)$ and $R_1(\tau)$.

7. Figure 2.4.24 shows part of the sequence of binary pulses produced by a pseudo-random generator. The basic pulses have amplitude 5 V and duration 1 μs and the complete sequence is repeated once per second. The succession of pulses in each sequence may be assumed to be random.

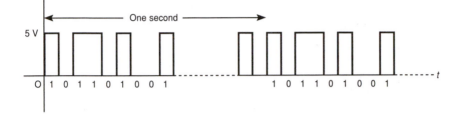

Figure 2.4.24 *Pseudo-random sequence*

Sketch the finite autocorrelation function for the basic sequence and deduce the corresponding energy spectrum. Hence sketch the power spectrum of the repeated sequence and estimate the root mean square voltage in a bandwidth from 1 Hz to 1 kHz.

To what extent can the waveform produced be regarded as equivalent to random noise?

8. The operations *I*, *T* and *T** are applied to a function $f(t)$.
   (i) Write down the values of TT^*, T^*T, TT, T^*T^*, TTT, $T^*T^*T^*$, $TTTT$, $T^*T^*T^*T^*$.
   (ii) Write down the values of TI, T^*I, $T-I$, T^*-I.

9.

2.5 Sampling and Quantisation

2.5.1 Introduction

There are many applications in Information Technology in which the signal of interest has to be sampled at discrete intervals. For example, in order to transmit several telephone conversations down a single channel, the signals may be sampled in turn at a suitable rate, the succession of samples transmitted in sequence, and the individual conversations reconstructed from the samples at the receiver. Similarly, if an analogue signal is to be processed by a computer, an 'instantaneous' value of the signal must be digitised (fed to an analogue-to-digital converter or ADC), the required processing carried out, and an analogue value produced at the output via a digital-to-analogue converter or DAC. Figure 2.5.1(a) and (b) shows such a system. For simplicity, the computer simply reads in the values and immediately puts them

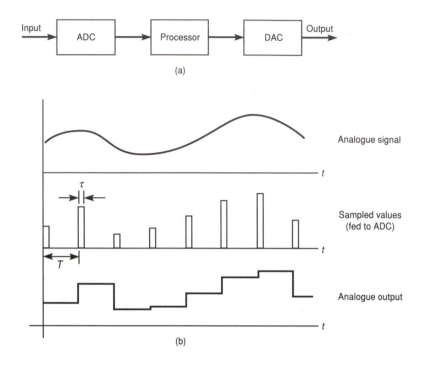

Figure 2.5.1 *(a) Digital processing system, (b) input and output signals*

out again, with a short time-delay. The output value from the DAC is held constant until updated at the next output.

In the system of Figure 2.5.1 two quite distinct processes are actually involved, namely sampling and quantisation. Sampling means taking the values of a signal at specific (usually equally spaced) instants in time, but the sampled values are still analogue and can have any value within the overall range of the signal. Quantisation, on the other hand, means rounding the signal values to the nearest discrete value in a set of allowed discrete values, for example to the nearest of the 256 levels in an eight-bit system. The two processes are sometimes said to be *orthogonal* since they are independent and may be done in any order. Figure 2.5.2 shows that the same result is obtained whichever process is done first.

Clearly we need to know what effects the sampling and quantisation processes have on the reconstructed signal, and how they are related to the sampling interval T, the sampling pulse width τ and the number of levels in the quantiser. We will discuss the sampling process first and will consider the effects of quantisation later in section 2.5.5.

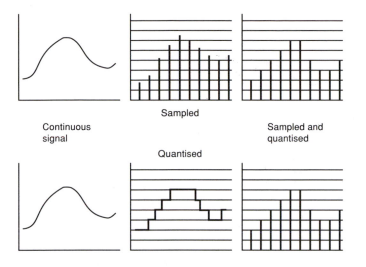

Figure 2.5.2 *Sampling and quantisation*

2.5.2 *The Spectrum of Sampled Signals*

The process of sampling an analogue signal is illustrated in Figure 2.5.3, in which a switch S is periodically connected to the signal for a short time τ.

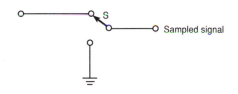

Figure 2.5.3 *The sampling process*

The sampling process can be thought of as the product between the signal $f(t)$ and a switching waveform $s(t)$, comprising narrow equally-spaced rectangular pulses of width τ as shown in Figure 2.5.4; the process is often referred to as 'natural sampling' for a switching waveform of this type. It follows, therefore, that the spectrum of the sampled waveform is the convolution between the spectrum of $f(t)$ and the spectrum of $s(t)$. The former has been assumed to be triangular for convenience, with bandwidth B, whereas the latter is a set of impulses with envelope determined by a sinc function (dependent on the pulse width τ).

The spectrum of the sampled signal $F_s(f)$ is just a series of repetitions of the original spectrum $F(f)$, spaced at intervals of $1/T$ and decreasing in height

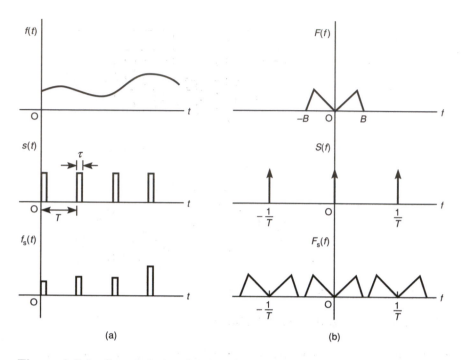

Figure 2.5.4 *Sampled signal (a) and its spectrum (b)*

according to the sinc function. It is clear that any one of these repetitions contains all the information in the original signal. For example, a low-pass filter of bandwidth a little more than B would extract the central section of $F_s(f)$ or a bandpass filter centred on $1/T$ (with similar bandwidth) would extract the first repeated section (strictly the pair at $1/T$ and $-1/T$).

It appears, therefore, that we can completely recover the original signal from the sampled version, without any loss of information at all! However, there is a critical restriction: the repeated spectra must not overlap with one another. This requires that the separation between spectra is greater than twice the bandwidth of the signal; that is

$$\frac{1}{T} \geq 2B$$

Strictly it is the highest frequency component that is important, since the bandwidth need not extend down to zero.

This is the well-known sampling theorem, usually stated as follows.

Sampling Theorem: If a signal contains no frequencies outside the range zero to B Hz, it is completely determined by the values of its ordinates spaced at less than or equal to $1/2B$ seconds.

Expressed slightly differently, if a signal has a bandwidth B we need to take at least $2B$ samples per second to retain all the information in the signal. If we take fewer than $2B$ samples/s the signal cannot be reconstructed, whereas if we take more its reconstruction is easier, since the tolerances on the filter required are less. This is illustrated in Figure 2.5.5.

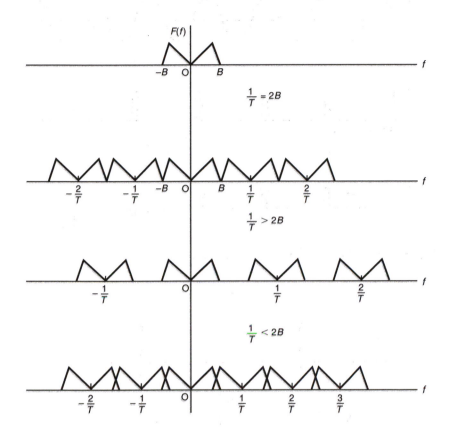

Figure 2.5.5 *Effect of sampling rate on spectrum*

When the sampling rate is insufficient, so that the repeated spectra overlap, an important effect known as *aliasing* occurs. This is the appearance of frequencies in the 'recovered' signal that were not in the original signal. This can be seen better by considering a sine wave, as in Figure 2.5.6. If the frequency is f_o it should be sampled at a rate of at least $2f_o$. However, the figure shows that sampling at intervals of slightly greater than T_o $(= 1/f_o)$ causes a low-frequency sine wave to appear. By considering the frequency domain representation as in Figure 2.5.7, we can see that we do indeed have an unwanted component at frequency $(1/T - 1/T_o)$ or $(f - f_o)$ where f is the sampling frequency. In practical systems, *anti-aliasing filters* are always used to prevent such effects from occurring, by limiting the signal bandwidth appropriately.

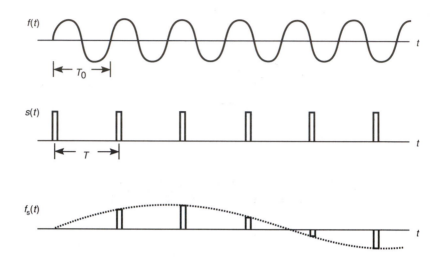

Figure 2.5.6   *The phenomenon of aliasing*

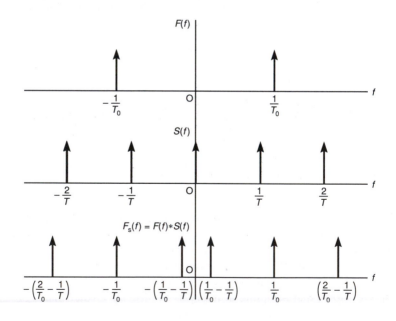

Figure 2.5.7   *Spectrum of 'aliased' signal (the convolution with the 2/T and −2/T components of S(f) have been omitted for clarity)*

The version of the sampling theorem discussed above is the most useful, but there are various extensions, for example to bandpass signals and also to 'sampling' in the frequency domain, to which a similar theorem applies, as one would expect.

2.5.3 Impulse Sampling and Sample/Hold

The sampling process (natural sampling) was illustrated above by means of a switch closing for a short time τ, producing a series of pulses proportional to the height of the signal and of width τ. Ideally we would like the pulse width to tend to zero, giving an 'instantaneous' sample, but the power in the pulse train would then tend to zero. Mathematically we can overcome the difficulty by considering that the sampling process is equivalent to multiplying the signal by a train of unit impulses, producing another train of impulses of strengths proportional to the signal values, as in Figure 2.5.8.

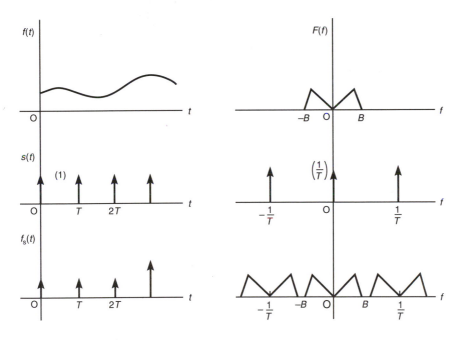

Figure 2.5.8 *Impulse sampling and its spectrum*

Mathematically, the process of taking a sample at a time τ is known as the sampling property of impulses (or sifting property), whereby the value of a function $f(t)$ at a time τ is given by

$$f(t) = \int_{-\infty}^{\infty} f(t)\delta(t - \tau)\mathrm{d}t$$

that is, multiplying $f(t)$ by a unit impulse at time τ. The periodic sampling by a series of impulses is denoted by a sampling train

$$s(t) = \sum_{n=-\infty}^{n=\infty} \delta(t - nT)$$

so the sampled waveform $f_s(t)$ is given by

$$f_s(t) = f(t) \sum_{n=-\infty}^{n=+\infty} \delta(t - nT)$$

The spectrum is similar to that of Figure 2.5.4, except that the repetitions are now all of the same height.

Unfortunately one cannot easily make a good approximation to an impulse, since a pulse of great height would be needed. However, the technique usually used is known as sample and hold (strictly the arrangement is known as 'zero-order hold'), whereby the 'instantaneous' value of one sample is held constant (for example, by an integrator) until the next sample is taken, as illustrated in Figure 2.5.9. A train of broad pulses is produced, as at the output of the DAC in Figure 2.5.1.

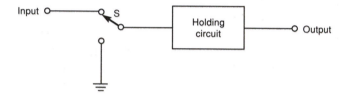

Figure 2.5.9 *Sample and hold circuit*

The spectrum of the sample-and-hold output can be obtained easily from the convolution theorems (surprise, surprise!) since the hold signal is just the convolution between the train of impulses (from the impulse sampler) and a single rectangular 'hold' pulse. The spectrum is therefore the product between the spectra of the impulse train (impulses) and the hold pulse (a sinc function), as shown in Figure 2.5.10.

We have: $f_s(t) = f(t) \times s(t)$ and $f_{sh}(t) = f_s(t) * h(t)$
Therefore: $F_s(f) = F(f) * S(f)$ and $F_{sh}(f) = F_s(f) \times H(f)$

The spectrum is slightly distorted by the sinc function due to the hold pulse, but it can be recovered completely (provided $1/T \geq 2B$) by a filter with the required gain characteristic (that is, a slight rise with frequency, up to B).

The concept of 'impulse sampling' is thus very useful, since although it is difficult to implement directly, it permits us to analyse the waveforms actually produced in practice.

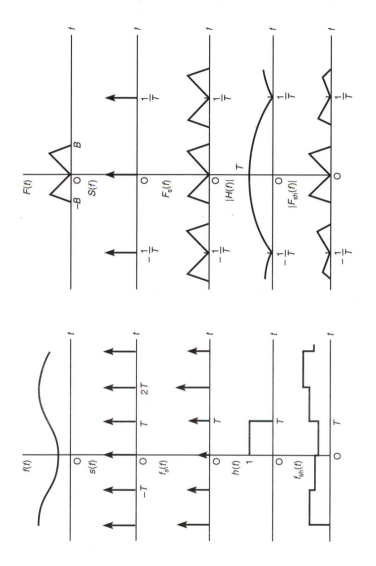

Figure 2.5.10 *Spectrum of sample-and-hold system*

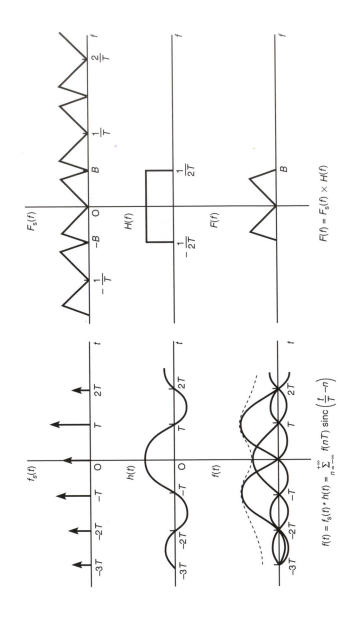

Figure 2.5.11 *Recovery of signal from samples*

2.5.4 Recovery of Signal from Samples

We have seen above that complete recovery of the signal from its samples is possible in principle provided that we have at least 2B samples per second. The result is really rather surprising, since one tends to think that the waveform of a signal between samples cannot possibly be known exactly. However, by transforming to the frequency domain we have seen that it must be completely recoverable at all times, that is, even between samples. In reconstructing the waveform in the time domain each value is determined in part by all the infinite set of impulses, and it is this, together with the fact that $f(t)$ is specifically limited in bandwidth (which limits its rate of change) that makes complete recovery possible.

$f(t)$ is recovered by a filter, as shown in Figure 2.5.11, where an ideal filter is assumed whose response falls immediately to zero at $1/2T$, and with no phase shift. Such a response cannot be obtained in practice, of course, so that sampling at a rate greater than 2B is necessary (typically 8000/s for a 3 kHz telephone channel) to ease the filter constraints.

The time equivalent of the ideal filter characteristic is a sinc function, with zeros at T, $2T$, $3T$ etc. (it is actually the impulse response of the filter, of course). Multiplying $F_s(f)$ by $H(f)$ gives $F(f)$, so in the time domain we obtain $f(t)$ by convolving the sinc function $h(t)$ with the impulses $f_s(t)$. This produces a set of sinc functions centred on 0, T, $2T$ etc., with heights governed by the appropriate sample values. For a given sample the sinc function has a peak height equal to its own sample value and is zero at all other sample points. $f(t)$ is thus reconstructed by adding sinc functions centred on all the sample values. In the particular case sketched, where the filter bandwidth is $1/2T$ ($= B$) the sinc functions determine only their own sample value, so that the samples are (just) independent of one another. However, with more rapid sampling this would no longer be the case.

In practice, therefore, one samples at a rate greater than 2B (having specifically limited the signal bandwidth to B by means of an anti-aliasing filter) so that a practical filter can be designed to recover $f(t)$. Its operation is equivalent to the addition of the sinc functions, fitting in the 'correct' values between samples and permitting a 'perfect' reconstruction of $f(t)$.

2.5.5 Quantisation

When a signal $f(t)$ is quantised before storage in a computer or before being transmitted in a communication system (as in Figure 2.5.2 above) and subsequently read out from the computer after processing or decoded in the receiver, the recovered signal (even in the absence of any errors) is, of course, simply the quantised signal, not the original signal before quantising. The difference between the two is known as *quantising noise* and is illustrated in Figure 2.5.12. It sets a limit to the signal-to-noise ratio obtainable in a digital system.

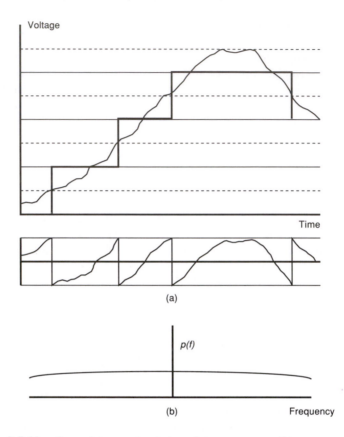

Figure 2.5.12 *Quantising noise (a) and its spectrum (b)*

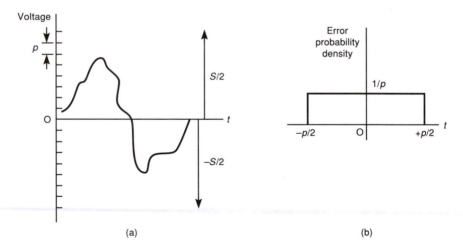

Figure 2.5.13 *Quantising errors (a) and their probability distribution (b)*

It is fairly simple to work out its magnitude. Consider an m-bit system as in Figure 2.5.13(a) with a maximum signal range a which is quantised into M $(= 2^m)$ levels so the spacing between levels is $p = a/M$.

Assuming that quantising errors may have any value between $+p/2$ and $-p/2$ with equal probability, the probability density distribution for errors is uniform as in Figure 2.5.13(b) and the mean square error $\overline{q^2}$ is given by:

$$\overline{q^2} = \int_{-p/2}^{p/2} \frac{1}{p} v^2 \, dv = \frac{p^2}{12}$$

The mean square signal may be found in a similar manner. Assuming that the signal ranges from $a/2$ to $-a/2$ and that all levels are equiprobable, the probability density distribution has a value of $1/((M-1)p)$ and extends from $-(M-1)p/2$ to $+(M-1)p/2$ (because of the zero level), and we find that $\overline{f^2} = (M-1)^2 \, p^2/12$. Since $M \gg 1$ in practice we have the simple result that the output signal-power-to-noise ratio is given by:

$$\left(\frac{S}{N}\right)_o \approx M^2$$

The result obtained depends on the exact assumptions made regarding the properties of the signal and it is sometimes given in a different form. The largest possible sine wave has peak-to-peak amplitude $= 2^m \times p$, so its mean square value is $1/2(2^m \times p/2)^2 = 2^{2m}p^2/8$ and the corresponding power-to-noise ratio is 1.5×2^{2m}; expressed in decibels this is $6.02m + 1.76$, a frequently quoted result Similarly, in a speech waveform the mean amplitude is much less than the peak and the result is often quoted as $3 * M^2/\alpha$, where α is the peak-to-mean power ratio, or as an r.m.s. ratio in dB of $4.7 + 6m - \alpha$ dB. Perhaps $(S/N) \approx M^2$ is easier to remember!

The spectrum of quantising noise is not easy to evaluate exactly. The signal consists of both slowly and rapidly changing sections, as in Figure 2.5.12, and has a sinc function shape. Fortunately, however, the exact shape does not matter, since the noise is clearly of much wider bandwidth than the signal and can simply be considered wideband (that is, white, within the signal bandwidth).

Quantising noise can be dramatically reduced by *oversampling*, that is, sampling at a rate greater than that required by the sampling theorem. The quantising noise is spread over a wide bandwidth and the noise within the signal bandwidth is therefore reduced. It can be shown that the r.m.s. quantising noise within the signal bandwidth is given by

$$\frac{(p^2 / 12)}{(OSR)^{1/2}}$$

where OSR is the oversampling ratio. Modern ADCs use oversampling ratios of 256 or more in conjunction with a feedback system (which compares the quantised output with the input) and can achieve resolutions as high as 24 bits (at fairly low frequencies).

2.5.6   Examples for Solution

1. A waveform is given by $f(t) = \frac{1}{2} \cos 3\omega_o t - \cos \omega_o t$. Simulate the sampling process by plotting the values of $f(t)$ at intervals of $\pi/3$ from zero to 2π (that is, sampling interval = period/6 = twice frequency of the third harmonic). Reconstruct the waveform by erecting sinc functions (by eye) at each sampling point, of maximum value equal to the sample value at that point, and zero at all other sampling points.
   Why is the reconstruction imperfect near zero and 2π?
2. The signal $f(t)$ band-limited to 2000 Hz is sketched in Figure 2.5.14(a) and the modulus of its spectrum in Figure 2.5.14(b). It is impulsively sampled at a rate of 5000 samples per second, and the sampled signal fed to a zero-order hold circuit of hold time 0.2 ms.
   (i) Sketch the impulsively sampled signal and its spectrum, and explain how $f(t)$ could be recovered.
   (ii) Sketch the output of the hold circuit and its spectrum, and explain how $f(t)$ could be recovered.
   (iii) Discuss the recovery of $f(t)$ when: (a) the sampling rate is reduced to 3000 samples/s and (b) the holding time is reduced to 0.1 ms (sampling rate 5000 samples/s).

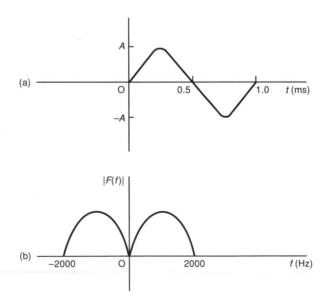

Figure 2.5.14   *Band-limited signal (a) and its spectrum (b)*

3   Coding

3.1   Introduction

The information produced by most sources of information is not in a suitable form for transmission via a channel, so that some type of coding procedure is necessary. The channel may be analogue or digital, and will have various characteristics such as bandwidth, transmitted power, error rate etc. Two important requirements of the system may be firstly that the information be transmitted as rapidly as possible and secondly that it be transmitted with as few errors as possible. These two requirements cannot easily be attained together; in order to increase the speed the redundancy of the code used has to be low, whereas to reduce the error rate the redundancy has to be high. However, we will see that Information Theory offers a surprising compromise, whereby under certain conditions both can be achieved together. The two requirements are often referred to as *source coding* and *channel coding* respectively.

3.2   Coding in Noiseless Channels: Source Coding

In this section we will consider only the first requirement, that of speed of transmission, and assume that we are dealing with a noiseless digital channel (that is, there are no transmission errors). The channel will be a binary channel (unless otherwise specified) and will have a certain fixed bandwidth B. The bandwidth of a channel limits the rate at which the signal can change, so that the number of pulses per second that can be sent down the channel is limited. As will be discussed in Chapter 4 the maximum pulse rate for a channel of bandwidth B (without unacceptable attenuation) is known as the Nyquist rate of $2B$ pulses/s. A telephone channel of bandwidth 3000 Hz can thus accept up to 6000 pulses/s. This means that the only way the speed of transmission via a given channel of information can be increased is by the choice of an efficient coding system, with minimum redundancy, so that fewer binary digits need to be transmitted.

3.2.1   Simple Binary Codes

We shall assume that we have a source of information, producing a source alphabet of symbols s_1, s_2 etc. of probabilities p_1, p_2 etc. One of the symbols may be a space symbol for separating groups of symbols. The symbols will be represented by a series of binary digits, the code for each symbol being known as a code word. A series of symbols, or the corresponding code words, will constitute a message. We shall deduce some of the important properties of binary codes by inspecting the five codes shown in Table 3.2.1 for a set of four source symbols.

Table 3.2.1 *Simple binary codes*

Code no.	s_1	s_2	s_3	s_4
1	0	10	110	11
2	00	01	10	11
3	0	10	110	1110
4	0	01	011	0111
5	0	10	110	111

It can be seen that code 1 has an ambiguity, in that the code for the sequence s_4, s_1 is the same as that for s_3. This code does not have the required property of being *uniquely decodable* and is therefore not usable without modification. Code 2 is a simple equal-length code; it should be noticed that, provided one starts at the beginning, no special space symbols are needed with this code since the binary digits can always be decoded in pairs. Code 3 is known as a comma code since the digit '0' is effectively used as a separator between code words. Such a system would clearly become very inefficient for a large source alphabet.

Codes 4 and 5 are both usable, but code 5 has an important property and is said to be *instantaneous*, whereas code 4 lacks this property. The distinction can be seen by considering the binary codes shown below for the sequence s_1, s_2, s_3, s_4, no space symbols being used.

        Code 4   0 0 1 0 1 1 0 1 1 1
        Code 5   0 1 0 1 1 0 1 1 1

With code 5, as soon as the code for a complete code word has been received, one can immediately decode that code word. However, for code 4, one has to look ahead at succeeding symbols before decoding is possible. An instantaneous code provides two important advantages: decoding is simple and no additional space symbols are required (provided that one starts at the beginning and that no errors occur). The required condition for an instantaneous code is that no code word be a *prefix* on another code word. Code 5 clearly satisfies this, whereas in code 4 each code word is a prefix of the succeeding one.

One may ask at this stage whether it would not be easier simply to use equal-length codes, such as code 2. The answer is that it is precisely by using codes of different lengths that we can get an improvement in the efficiency of coding. All that has to be done is to represent the most probable symbols by the shortest code words, and use longer words for the less probable symbols. This is done to some extent in Morse Code, where the most probable letter 'E' is represented by a single dot. Such codes are said to be *compact*.

Suppose the probabilities of the four symbols s_1, s_2, s_3, s_4 above are 0.6, 0.2, 0.1 and 0.1 respectively. A message of ten symbols would require 20 binary digits using the equal-length code 2. Using code 5 the message would require

$$10\ [(0.6 \times 1) + (0.2 \times 2) + (0.1 \times 3) + (0.1 \times 3)] = 16 \text{ binary digits}$$

The *average length L* of a code is an important parameter in determining the efficiency of a system. It is given by $L = \Sigma\, p_i\, l_i$ where l_i is the length (in binary digits) of the ith code word of probability p_i. In the example above, L is the expression in square brackets and has a value of 1.6 binary digits per symbol for the compact code.

The *efficiency* of coding E is defined by $E = H/L$ and is always less than or equal to unity. In the above example $H = 1.571$ so $E = 0.98$ for the compact code but is only 0.79 for the equal-length code (which has $L = 2$).

3.2.2 Compact Instantaneous Codes

The desirable properties of a code are clearly that it be uniquely decodable, instantaneous and compact, and we therefore need a method of finding such codes. This is easy by trial and error in simple cases, but becomes very complicated for large alphabets.

There is an interesting special case, in which the symbol probabilities p_i are of the form $p_i = (1/2)^{\alpha_i}$ where α_i is integral. It can be shown (Abramson, 1963) that perfect coding is obtained by setting the lengths of code words $l_i = \alpha_i$.

EXAMPLE 3.2.1

Four symbols s_1, s_2, s_3, s_4 have probabilities $^1/_2$, $^1/_4$, $^1/_8$, $^1/_8$ respectively. Find an instantaneous compact code, and deduce its average length and efficiency.

Using the relation above, $l_1 = 1$, $l_2 = 2$, $l_3 = l_4 = 3$. A suitable code would be $s_1 = 0$, $s_2 = 10$, $s_3 = 110$, $s_4 = 111$.

The average length is given by

$$L = \sum p_i\, l_i = (1 \times {}^1/_2 + 2 \times {}^1/_4 + 3 \times {}^1/_8 + 3 \times {}^1/_8)$$

$$= 1^3/_4 \text{ binary digits}$$

The source entropy is

$$H = -\sum p_i \log p_i = -({}^1/_2 \log {}^1/_2 + {}^1/_4 \log {}^1/_4 + 2 \times {}^1/_8 \log {}^1/_8)$$

$$= 1^3/_4 \text{ bits per symbol,}$$

numerically equal to the average length. This is not a coincidence. If we consider the stream of 0s and 1s representing a message, this stream carries most information (1 bit/digit) when the 0s and 1s are equiprobable. If the average length is L binary digits then on average each symbol gives rise to L bits of information. This is the best that can be done. If the 0s and 1s are not equiprobable, each symbol will produce on average more binary digits than are strictly necessary. The coding procedure can thus be seen to be essentially a matter of arranging that equal numbers of 0s and 1s are produced in the coded version of the message.

The efficiency H/L is exactly unity in this example and can always be made so if the probabilities are integral powers of one-half. The same process can be applied to codes of any base n if $p_i = (l/n)^{\alpha_i}$.

3.2.3   Coding Methods

Fano–Shannon Method

Although the above special case gives the required length of code words it does not actually produce the codes, though this is relatively easy if the lengths are known. A method of coding for any source was suggested by Shannon and Weaver (1949) and modified by Fano (1961). It involves writing the symbol probabilities in a table in descending order, and dividing them into pairs, as shown in the example below.

EXAMPLE 3.2.2

Find a code using Fano–Shannon coding for a source of five symbols of probabilities 0.5, 0.2, 0.1, 0.1, 0.1. Find the efficiency.

Dividing lines are inserted to successively divide the probabilities into halves, quarters etc. (or as nearly as possible). A '0' and '1' are added to the code at each division and the final code obtained by reading from the right towards each symbol, writing down the appropriate sequence of 0s and 1s.

						code
s_1	0.5			0		0
s_2	0.2	0				100
			0			
s_3	0.1	1				101
s_4	0.1	0		1		110
			1			
s_5	0.1	1				111

$L = 0.5 \times 1 + 0.2 \times 3 + 3 \times 0.1 \times 3 = 2.0$

$H = 1.96$
$E = 0.98$

This coding procedure is very straightforward and can be applied easily to large source alphabets (see section 3.2.6, exercise 4). It is easy to implement by a computer program.

Huffman's Method

A rather more efficient scheme was developed by Huffman (1952), in which the probabilities are again placed in descending order but added in pairs from the bottom and reordered, as shown in example 3.2.3.

EXAMPLE 3.2.3

Use Huffman's method for the code in the example above.

$L = 0.5 \times 1 + 0.2 \times 2 + 0.1 \times 3 + 2 \times 0.1 \times 4 = 2.0$
$H = 1.96$
$E = 0.98$

A '0' or '1' is placed at each branch, and the code obtained by reading from the right towards the symbol required.

The code obtained is different from Fano–Shannon's but the efficiency turns out to be the same. The inherent improvement, due to the reordering, is only small and appears only in complicated systems. Huffman's method uses rather a lot of paper in such cases, though a computer solution is again easy to implement.

3.2.4 Shannon's First Theorem

Both the Fano–Shannon and Huffman methods usually produce efficient codes. However, a further increase in efficiency can be obtained by grouping the source symbols in pairs, threes etc., and applying the coding procedure to the relevant probabilities of the group chosen. If the symbols are independent (as assumed above) the probabilities of the groups are just the products of the single probabilities, but of course if intersymbol influence is present then the joint probabilities will reflect this.

Shannon proved formally that if the source symbols are coded in groups of n, then the average length per single symbol tends to the source entropy H as n tends to infinity. That is

$$\lim_{n \to \infty} \frac{L_n}{n} = H$$

where L_n is the average length of the encoding of a group of n source symbols.

As before, the essence of the idea is that the coding procedure tends to produce equal numbers of 0s and 1s at the output, grouping in large groups simply making this more achievable. The coding process is sometimes known as 'matching the source to the channel', that is, making the output of the coder as suitable as possible for the channel used.

EXAMPLE 3.2.4

An information source produces a long sequence of three independent symbols A, B, C with probabilities 16/20, 3/20 and 1/20 respectively; 100 such symbols are produced per second. The information is to be transmitted via a noiseless binary channel which can transmit up to 100 binary digits per second. Design a suitable compact instantaneous code and find the probabilities of the binary digits produced.

It is a good idea to check whether the problem can be solved by finding the entropy of the source: $H = - \Sigma\, p_i \log p_i = 0.884$ bits so the source rate is 88.4 bits/s. This is less than the channel capacity, which is 100 bits/s since one binary digit can carry a maximum of 1 bit of information. Coding singly, using the Fano–Shannon method

A	16/20		0	0
B	3/20	0		10
C	1/20	1	1	11

$L = 1.2$ bits/symbol and the source rate $= 120$ binary digits/s. This is too great, so coding in pairs is necessary. As seen from the coding tree at the top of page 99, $L = 1.865$ per pair $= 0.9325$ bits/symbol.

The source rate is now less than 100 and the efficiency $E = 93$ per cent.

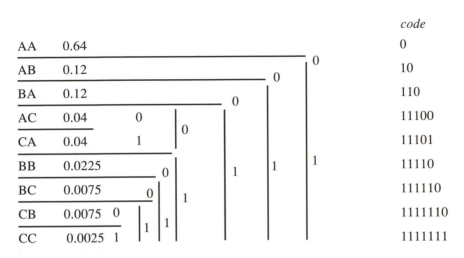

code

AA	0.64	0
AB	0.12	10
BA	0.12	110
AC	0.04	11100
CA	0.04	11101
BB	0.0225	11110
BC	0.0075	111110
CB	0.0075	1111110
CC	0.0025	1111111

The probabilities of 0 and 1 in the coded stream are given by

$$\frac{p(0)}{p(1)} = \frac{0.64 \times 1 + 0.12 \times 2 + 0.04 \times 3 + 0.02 \times 1 + 0.01 \times 2}{0.12 \times 3 + 0.04 \times 7 + 0.02 \times 4 + 0.01 \times 11 + 0.0025 \times 7}$$

$$= \frac{1.03}{0.853}$$

Thus $p(0) = 0.547$, $p(1) = 0.453$.

The entropy of the output stream (viewed as an independent source) is $-(p(0) \log p(0) + p(1) \log p(1)) = 0.993$ bits, which is very close to the maximum value of 1 bit which would be achieved for $p(0) = p(1)$.

3.2.5 *Applications of Compact Codes for Data Compression*

The growing importance of transmitting and storing large amounts of information in digital form has focused attention on the need to represent that information in as efficient a way as possible. However, in many cases the raw data use representations which are not designed for efficiency but for convenience. For example, the most common way for a computer system to store character-based files is by using the ASCII code, which uses 7 binary bits for every character. As has been shown previously, each character in the English language carries only 4.7 bits (for equiprobable characters) and considerably less if the probabilities are taken into account. Even allowing for the need to represent numerals, punctuation marks and both upper and lower case letters, ASCII is inefficient because no allowance is made for the frequency of occurrence of each character. Hence many systems concerned with transmitting or storing large amounts of character-based information employ source coding techniques to increase the efficiency of the representation. These coding techniques are often called *data compression codes* as they give the appearance of compressing the amount of data needed to be sent or stored.

In applications where it is vital that all the information contained in the message should be preserved through the compression and subsequent decompression stages, the coding methods used are referred to as *lossless* data compression. Other terms used are *reversible* and *noiseless* data compression methods. Applications, for example where digitised speech or visual information (such as still pictures or video) are to be transmitted or stored, are becoming more common. A television signal is transmitted at a rate of 25 frames per second (in the UK). If the analogue signal were to be digitised using a square array of 625 x 625 picture elements (pixels) and using 8 bits per pixel for each of the three primary colours, then the total information rate would become:

$$625 \times 625 \times 8 \times 3 \times 25 = 234 \times 10^6 \text{ bits/s}$$

which is clearly impractical for most transmission or storage purposes. However, much of the data in a digitised picture is redundant. It is likely that most parts of a frame will remain the same when compared with the next. If the changes between frames are transmitted rather than the frames themselves then a substantial saving in data rate can be made.

It is possible to make even greater savings in data rate if the quality of the image (or speech) data is allowed to deteriorate. For example, tests have shown that digitised speech can still be understood if all of the harmonic information is removed, allowing the data rate needed to convey a spoken message to be substantially reduced. However, it is the harmonics that allow the listener to identify the speaker and provide intonation to the speech. It is here that qualitative judgements have to be made as to the amount of data rate reduction that is acceptable. Is it desirable only to get the raw information across (that is, the spoken words) or must extra data be sent to provide interest and speaker identification? The coding mechanisms that can be used to reduce the amount of data needed to convey spoken or visual messages do not preserve all the original information, hence they are commonly called *lossy* data compression methods. Other terms used are *irreversible* and *noisy* data compression. The amount of information lost versus the data rate reduction achieved is a subjective trade-off.

3.2.5.1 Lossless Data Compression

Lossless coding methods fall into two distinct classes, those based on a dictionary approach and those based on a statistical approach. Dictionary (or substitution) codes are more widely used as they are usually easier (and hence quicker) to encode and decode, although statistical methods based on arithmetic codes offer greater potential for data reduction.

Dictionary-based Data Compression Codes

These codes rely on both the sender and the receiver of the information maintaining a dictionary of code strings which have occurred before in the information being transmitted. It is then only necessary to transmit a pointer to the location in the dictionary rather than the string itself. The various codes differ in the way the dictionary is built up and maintained. If the frequency of occurrence of letter strings in the data can be reliably predicted (for example, if the text is standard English) then a *static* dictionary can be used. Both sender and receiver store this dictionary as a look-up table and the message is transmitted as a series of pointers into the table. However, if the data do not conform to the predicted frequencies then the pointers will, quite often, be for single characters and no data reduction will occur. In fact, it is probable that an expansion in the data would occur as the pointers are likely to be longer (in bits) than the individual characters.

An obvious way of coding text-based data for maximum efficiency is to use a Huffman or a Shannon–Fano code, as described in section 3.2. However, these codes suffer the disadvantage of needing to know in advance the probability of occurrence of each symbol in the chosen alphabet. This can be predicted with reasonable certainty if it is known that the text will, for example, always be non-technical English, or always Polish, but not if either of these (or any other) language could be encountered.

In most practical applications of dictionary-based coding, adaptive techniques to rebuild the dictionary for every new data stream are used. This ensures that the compression is as close to optimal as possible but means that the encoding and decoding steps take more time. The best methods are all based on so-called *Ziv–Lempel* (also known as *Lempel–Ziv* or *LZ*) compression, which works by replacing incoming character strings or phrases by pointers to where they have previously occurred. Since the first exposition of the basic algorithm in 1977 (now known as *LZ77*) there have been many variants proposed, allowing different sizes of dictionary and of the maximum string length to be encoded. Some variants also use Huffman or Shannon–Fano coding for the pointers.

In 1978 Ziv and Lempel introduced another algorithm which has become known as *LZ78*. This worked on the basis of building up a dictionary character by character rather than phrase by phrase. LZ78 has also spawned many variants, the best known of which is described in detail here. It is known as *Lempel–Ziv–Welch (LZW)* compression and was originally proposed as a mechanism for compressing data to be stored on a fixed disk drive. The dictionary starts by containing pointers to all single characters in the alphabet to be compressed. The encoder then follows the algorithm shown below, where C is the next character in the incoming data and s is the longest string that has been seen before.

```
set s = null
do while C exists
read C
if sC exists in the dictionary
          then s = sC
else
          output pointer for s
          add sC to the dictionary
          s = C
loop
```

Taking the well known phrase 'to be or not to be', this would give the result shown in Table 3.2.2. Note that the space character is shown as \ and that the dictionary starts by containing all lower case letters (pointers 1 to 26) and the space character (code 27). The incoming character column shows the longest string to be matched in the dictionary.

Table 3.2.2 *LZW compression. Phrase to be compressed:*
to\be\or\not\to\be

Incoming character C	Pointer output	Added to dictionary sC
t	20	
o	15	to (28)
\	27	o\ (29)
b	2	\b (30)
e	5	be (31)
\	27	e\ (32)
o	15	\o (33)
r	18	or (34)
\	27	r\ (35)
n	14	\n (36)
o	15	no (37)
t	20	ot (38)
\	27	t\ (39)
to	28	\to (40)
\b	30	o\b (41)
e\	32	be\ (42)

The original implementation of LZW encoding specified a dictionary limited to 4096 pointers, with the first 255 locations being used for the normal ASCII characters and common extended ASCII characters.

The decoding of the data stream is accomplished in much the same way as the encoding – hence the dictionary is rebuilt at the receiver without having to be transmitted explicitly. This is the key advantage of all the

variants of LZ compression – the lack of need to send the dictionary between the transmitter and receiver. Most data compression methods which
have found widespread (such as PKZIP, LHA, ARJ, Compress, V.42$_{bis}$) use
variations of the LZ78 algorithm. Compression ratios of between 2 and 4:1
can be expected, depending on the precise nature of the data to be compressed.

Statistical Methods

Arithmetic Text Compression

Arithmetic coding does not attempt to assign specific codes or pointers to
the data to be compressed, rather it replaces the whole message with a single
number. Longer messages result in a number of greater precision. Like Huffman
coding, a precise knowledge of the probability of occurrence of any piece
of data is needed before coding can be performed. The application of arithmetic coding is best illustrated by an example. Take the string 'to be or n'
(*note*: a string of total length 10 characters has been chosen to help clarify
the explanation). The probability of occurrence of each character in this
string is

Space	0.3
b	0.1
e	0.1
n	0.1
o	0.2
r	0.1
t	0.1

These probabilities are then assigned a range within the total space 0 to 1.
It is not important in what order the characters are used:

Character	Low end	High end
Space	0	0.3
b	0.3	0.4
e	0.4	0.5
n	0.5	0.6
o	0.6	0.8
r	0.8	0.9
t	0.9	1

The string is encoded by successively considering the interval to which the
next character corresponds. The first character is a 't' , so the output number
must lie between 0.9 and 1.0. The next character is a 'o', so the range in
which the output number can fall must be narrowed to between 0.6 and 0.8
of the opening range. This process is continued for each character until the
whole string has been encoded:

t	0.9	1.0
o	0.96	0.98
space	0.960	0.966
b	0.9618	0.9624
e	0.96204	0.9621
space	0.962040	0.962058
o	0.9620508	0.9620544
r	0.96205368	0.96205404
space	0.962053680	0.962053788
n	0.9620537340	0.9620537344

The number 0.9620537340 uniquely encodes the message string 'to be or n'.

The decoding process is just a reverse of the encoding process. In this example, as the final number lies between 0.9 and 1, the first character must be a 't'. The next step is to subtract the low end of the range from the number and divide what is left by the width of the preceding range (0.1). This leaves 0.62053734, which lies between 0.6 and 0.8. Hence the next character must be 'o'. Subtracting 0.6 and dividing by 0.2 gives 0.1026867, which is between 0 and 0.3, so the next character is a 'space'. This recursive process uniquely decodes the correct string.

Practical applications of Arithmetic Coding are not widespread. The need for an accurate statistical model of the data, combined with the difficulty of representing such precise numbers in a conventional computer, are obvious limitations when compared with adaptive dictionary-based schemes. However, for some types of data they offer much greater compression ratios, so Arithmetic Coding has been used for some applications.

Run Length Encoding (RLE)
If the data to be compressed are in binary form, it can be advantageous to encode the data as sequence lengths. For example, the data 111100110000 could be represented as 14225. The starting code is indicated (1 or 0) followed by a number indicating the sequence length before it changes. As there are only two possible states, the next number must indicate a sequence of 0s. The most widespread use of RLE is in FAX machines. The data to be transmitted are naturally binary, 1 representing a black dot and 0 a white dot. The most likely sequence lengths are predicted and the numbers representing these are encoded using a Huffman tree. Other codes are used to indicate blank lines. FAX machines can achieve data compression ratios up to 20:1 using this method, saving a large amount of transmission time. FAX is discussed in more detail in Chapter 5.

3.2.5.2 Lossy Data Compression

The most common method used to achieve lossy compression is to smooth out rapidly changing parts of the data, thus lowering the required number of samples. This is achieved by transforming the data (usually an image or a sound) to the frequency domain and then ignoring the coefficients of the higher frequencies, which represent the rapid changes. The transform is usually the *Discrete Cosine Transform* (DCT) because this can be performed in software. How many of the coefficients are ignored will determine the compression ratio and the subjective loss of quality.

The need to transmit images and video signals over channels with limited bandwidth led to the formation of two groups charged with defining standards. These are the *Joint Picture Expert Group (JPEG)*, whose primary concern is with still images, and the *Motion Picture Expert Group (MPEG)*, whose main concern is with video images. Both these groups have produced compression standards which combine both lossy and lossless coding methods.

The JPEG standard starts by dividing the image into blocks of 8 × 8 picture elements (pixels). These are transformed using a DCT. The coefficients of the transformed image are then quantised, so that any small coefficients are reduced to zero. In most images the high frequency components are very small. The JPEG standard allows the quantisation step to be varied, allowing a trade-off between compression ratio and quality. After quantisation the coefficients are further encoded using a combination of Huffman and RLE methods. Compression ratios of 10:1 can be achieved with little loss to perceived quality, whilst ratios up to 100:1 can be defined, if the resultant image can still convey the essential information.

The MPEG standard makes use of the nature of a video image by assuming that most of the image will not change between frames. An outline description of the algorithm, omitting some of the steps, will give a flavour of the complexity of the processing needed. MPEG defines a series of frames to be reconstructed, thus it is better described from the point of view of the decompressor. Some frames, called 'I' or *Intra* frames arrive intact; they are just coded as still images. These frames are intended to give a known starting point and are transmitted once every 12 frame intervals. The 'P' or *Predicted* frames are generated from the most recently reconstructed I or P frame, using difference coefficients, generated by the encoder. Then 'B' or *Bi-directional* frames are generated using information from the closest two I or P frames. The sequence of frames at the decompressor is then:

IBBPBBPBBPBBIBBPBBP . . .

Each frame, or set of difference coefficients, is encoded in a similar manner to the JPEG algorithm described above. Hence there are two places where information is lost – during the decomposition into I frames and difference coefficients, and during the quantisation of the DCT coefficients. Compression ratios up to 200:1 have been claimed for the MPEG algorithm. The different standards (MPEG I and MPEG II) relate to different starting points in terms of pixels per frame.

Much research has gone into new forms of data compression for images, specifically into techniques involving more efficient transforms. *Wavelet* transforms are claimed to offer a factor of 2 improvement in compression ratio over techniques which use DCT, but are more difficult to calculate. *Fractal* compression attempts to divide areas of the image up into small blocks which can then be replicated to form the entire image; it relies on the complex mathematics of chaos theory and an explanation is beyond the scope of this book.

3.2.6   Examples for Solution

1. Construct a compact instantaneous code for five source symbols of probabilities 1/2, 1/4, 1/8, 1/16, 1/16. Show that the average code length is equal to the source entropy.
2. The types and numbers of vehicles passing a point in a road are to be recorded automatically on a tape recorder. A binary code is to be assigned to each type of vehicle and the appropriate code recorded on the passage of that type. The average numbers of vehicles per hour are as follows:

   Cars: 500   Motorcycles: 50   Buses: 25   Lorries: 200   Mopeds: 50
   Vans: 100   Cycles: 50   Others: 25

   Design a suitable compact instantaneous code. Find its efficiency and compare it with that of a simple equal-length binary code. Comment on the feasibility and usefulness of this system.
3. A source produces three symbols A, B, C with probabilities 0.6, 0.3, 0.1. Find a suitable binary code using Huffman's method:
   (a) using the source symbols separately
   (b) using the source symbols in pairs.
   Find the average code length and efficiency in each case. Find the probabilities of the 0s and 1s in the output stream, and the entropy of the coded output.
4. Use Fano's procedure to find a compact code for the letters of the alphabet, using the probabilities below, and estimate the saving in memory compared with standard ASCII code.

% occurrence	Letter	% occurrence	Letter
18	space	5	N, R
11	E	3	D, L
8	T	2	C, F, M, U
7	A	1	B, G, K, V, W, Y, P
6	H, I, O, S	0.25	J, Q, X, Z

   Note: these values have been rounded from the accepted probabilities.

5. Use the LZW algorithm outlined in section 3.2.5.1 to encode the data string 'dabbadabbada'.

6. Use Arithmetic Coding to encode the data string 'dabbadabbada' and compare the compression achieved with your answer to example 5.

3.3 Coding in Noisy Channels: Channel Coding

3.3.1 Introduction

We have seen previously that in a communication system we may be concerned either with the speed of transmission of information, or with the accuracy of transmission. In section 3.2 we considered speed of transmission in noiseless systems, and noted that the efficiency of coding was a measure of the redundancy of the system, perfect efficiency implying zero redundancy. In this section we will consider the accuracy of transmission in a noisy system. Clearly, redundancy will have to be added to give protection against errors. For example, in a system with two input symbols A and B, coding A = 000 and B = 111 gives a reduction in the overall error rate (compared with A = 0, B = 1), since single errors can be tolerated (if a majority decision is used for decoding, as in section 1.1.4). Unfortunately the redundancy is greatly increased, and it appears at first sight that there is bound to be a simple exchange between error rate and redundancy. However, it will be shown that Information Theory predicts that the situation is not as bad as this, and that, subject to certain restrictions, a low error rate can be achieved together with a high data rate (or low redundancy).

We will assume that we are dealing with a binary symmetric channel with binary error probability p. The symbols to be transmitted will be coded as groups of binary digits (code words) and we will need to evaluate the symbol error probability $p(e)$ for a given coding system. It is important to distinguish this from the binary error probability; in the above example $p(e)$ is the sum of the probabilities of two and three binary errors in a sequence of three binary digits, whereas the binary error probability is simply the probability that a given binary digit is received in error.

The information rate R is the information per digit produced by the source before transmission, so in the above example $R = 1/3$ bit per binary digit (since three digits are used to carry one bit of information). The relation between R and $p(e)$ will be our main concern in this section.

In order to find $p(e)$ we need to find the probability of getting a given number of errors in a given number of binary digits. The binomial distribution, discussed in Appendix 1, provides precisely what we require. It refers to a process with two outcomes of probabilities p (for an error, say) and $\bar{p}$ (for no error), and states that the probability $p(r)$ of getting r errors in n trials is given by the $(r + 1)$th term of the binomial expansion of $(p + \bar{p})^n$; that is

$$p(r) = \frac{n!}{(n-r)! \; r!} p^r \bar{p}^{(n-r)}$$

EXAMPLE 3.3.1 BINARY CODING FOR ERROR PROTECTION

We will assume that we are coding symbols for transmission in a binary symmetric channel of binary error probability $p = 0.01$, and will use code words that are five binary digits in length in all cases.

Coding by repetition
Suppose we code for two symbols A and B only, taking A = 00000, B = 11111. It is necessary to interpret received symbols according to a decision rule fixed in advance; in this case we will use a simple majority rule, so that the received sequence is interpreted as A if it has more 0s than 1s and vice versa. Two errors can be tolerated without producing a symbol error and $p(e)$ is given by the sum of the probabilities of three, four and five errors in five trials:

$$(\bar{p} + p)^5 = \bar{p}^5 + 5\bar{p}^4 p + 10\bar{p}^3 p^2 + 10\bar{p}^2 p^3 + 5\bar{p}^4 p + p^5$$
$$p(r) \qquad p(0) \;\; p(1) \qquad p(2) \qquad p(3) \qquad p(4) \qquad p(5)$$

$$p(e) = p(3) + p(4) + p(5) = 9.8 \times 10^{-6}$$

The information rate $R = 1/5$ bits/binary digit. Clearly, we can decrease $p(e)$ only at the expense of decreasing R; if we used 7 binary digits we would find $p(e) = 3.5 \times 10^{-9}$ and $R = 1/7$.

Coding by selection of code words
A more powerful and general approach is to select only certain code words for use out of all the possible code words. Using five digits there are 32 possible code words, as shown in Table 3.3.1. We can view the repetition method above as a selection of only two of the possible code words.

(i) Two selections (that is, repetition)

A = 00000, B = 11111

This gives $p(e) = 9.8 \; 10^{-6}$, $R = 1/5$ as above.

(ii) Thirty-two selections
At the other extreme we could elect to use all 32 code words (including some punctuation besides the 26 upper case letters)

A = 00000, B = 00001, C = 00010, ... etc.

Since all combinations are used, a single binary error will produce a symbol error. Therefore

$$p(e) = p(1) + p(2) + p(3) + p(4) + p(5) = 1 - p(0) = 0.049$$

Table 3.3.1 *Coding by selection of code words*

	32 selections			2 selections		4 selections	
A	00000	Q	10000	A	00000	A	00000
B	00001	R	10001	B	11111	B	00111
C	00010	S	10010			C	11001
D	00011	T	10011			D	11110
E	00100	U	10100				
F	00101	V	10101				
G	00110	W	10110				
H	00111	X	10111				
I	01000	Y	11000				
J	01001	Z	11001				
K	01010		11010				
L	01011	.	11011				
M	01100	,	11100				
N	01101	;	11101				
O	01110	:	11110				
P	01111	?	11111				

The information rate $R = 1$ bit/binary digit (assuming 32 equiprobable symbols, each requiring 5 digits).

(iii) Four selections

A compromise between these two extremes is possible, by selecting a number between two and thirty-two. To obtain a reasonable data rate we want to use a lot of code words, but to keep the symbol error rate low those selected should be as different from one another as possible. Consider the selection:

$$A = 00000, B = 00111, C = 11001, D = 11110$$

Every code word differs from all the others in at least three digit positions. There are many other possible choices, but only four such code words are possible in each case. The number of digit positions in which two code words differ is known as the *Hamming Distance*, and the least difference between any members of a given set of code words is called the *Minimum Hamming Distance*, d_{min}. It is clear that single errors can be tolerated, since a received code word with one error will still be closer to the transmitted code word than to any of the other selected code words. However, symbol errors will occur for two or more errors. Therefore

$$p(e) = p(2) + p(3) + p(4) + p(5) \approx p(2) = 9.7 \times 10^{-4}$$

The information rate $R = 2/5$ bits/binary digit.

A comparison of the three cases is given in Table 3.3.2.

Table 3.3.2

No. of selections	$p(e)$	R
2	9.8×10^{-6}	0.2
4	9.7×10^{-4}	0.4
32	0.049	1

3.3.2 Shannon's Second Theorem

The example above suggests that a compromise between low error rate and high data rate may be possible by using long groups of digits (more than five, say) so that there are a large number of possible code words, and by selecting a sufficient number for use to keep the R values reasonable but with the codewords sufficiently different from one another to make $p(e)$ acceptably low. This is the principle of Shannon's second theorem. Shannon showed (see Shannon and Weaver, 1949) that if coding is in long groups of n binary digits, then provided that the number of selected groups M is less than or equal to 2^{nC}, where C is the channel capacity, the symbol error rate $p(e)$ can be made arbitrarily small as n tends to infinity. That is

$$p(e) \rightarrow 0 \text{ as } n \rightarrow \infty \text{ provided } M \leq 2^{nC}$$

Assuming that the M selected code words are equiprobable, the information rate $R = (\log M)/n$ bits/symbol. However, since $(\log 2^{nC})/n = C$ the requirement is equivalent to $R \leq C$. In other words, provided that $R \leq C$ (that is, $R/C \leq 1$) then $p(e)$ tends to zero as n tends to infinity. We thus have the surprising result that, irrespective of the binary error probability, information can be transmitted at any rate up to and including the channel capacity with *no net errors* (that is, symbol errors in the decoded output).

Applying this to the system discussed above, the capacity C with $p = 0.01$ is 0.919 bits. Choosing 2 or 4 code words gives $R/C = 0.22$ or 0.44 respectively, whereas choosing 32 code words gives $R/C > 1$ (since $R = 1$). However, $p(e)$ does not tend to zero, even for 2 or 4 code words, since n is not sufficiently large. Of course, what the theorem does not tell us is how large the group of binary digits needs to be to give a specified symbol error rate. Unfortunately n has to be rather large in practice, as can be seen from the examples below.

EXAMPLE 3.3.2

Suppose $n = 10$ (with $p = 0.01$ and $C = 0.919$ as above). The total number of possible combinations is $2^n = 1024$. The number of selected code words

M must be less than $2^{nC} \approx 2^9$, so choosing $M = 2^9$ (so that $R/C \approx 1$) we would be using 512 symbols or 1 in 2 of the possible code words as symbols. There would be no way of selecting symbols so as to get a low error rate. On the other hand, if we chose $M = 32$ (so $R = 5/10 = 0.5$ and $R/C = 0.5/0.919 = 0.54$) we would have a choice of 1 in 32 of code words, which could be chosen to differ from one another in several digit positions and produce a reasonably low error rate.

Suppose now that $n = 100$; the total number of combinations is now 2^{100} and if we choose $R = C$ we have $M \approx 2^{90}$. Only about 1 in 1000 of the possible code words are used, so they can be chosen to give a very low error rate. Choosing $R \ll C$ would make the error rate even better. Unfortunately, $M \approx 2^{90}$ (or proportionately less if $R \ll C$)) is considerably more than the vocabulary of even the most ardent player of that unmentionable word game, so such a system is totally impracticable. However, various compromises are possible; for example, taking $n = 20$ and choosing $M = 64$ gives $R/C = 1/3$ with a selection of only 1 in 2^{14} combinations as code words.

Shannon's theorem points us in the right direction, showing what the theoretical limit is and indicating that the error rate falls as n increases to an extent depending on R/C. However, it does not help much in finding suitable codes and its direct application is not practical. It is nevertheless of great importance and has led to considerable developments in coding theory. Some of the codes developed will be outlined later in this chapter, but we will next consider some practical coding systems, viewing them in the light of Shannon's theorem.

3.3.3 Practical Codes for Error Detection and Correction

It is important to distinguish between error detection and error correction. Suppose we have a set of code words with a d_{min} of two. A single binary error will produce a word not in the list of those selected, so that the presence of a (binary) error will be detected, but there will be no way of knowing which code word was actually transmitted. However, if a set of code words has a d_{min} of three, a single error will produce a word nearest to the 'correct' code word (as seen above) so that single errors can be said to have been corrected. In general

to detect n errors $d_{min} \geq n + 1$

to correct n errors $d_{min} \geq 2n + 1$

A practical code must be constructed in such a way that any error introduced by the physical process of transmission involves an allowable code sequence changing to a non-allowable code sequence. If the bits in error are such that another allowable code sequence is produced by the changes then the code has broken down and incorrect data will be passed on. This important point must be borne in mind when specifying a code for a particular application. The probability and consequences of a code breaking down must be calculated and any remedial action specified. For example, the higher layers of a data communications system should be able to detect that an unrecoverable error has occurred.

Errors in long strings of data bits can generally be described either as random, implying that there is no correlation between the bit positions in error, or as burst, meaning that the bits in error are contained in a sequence. The burst error is a characteristic of many telecommunications applications. In a high speed communications channel changes introduced by electromagnetic impulses are likely to affect a group of neighbouring bits (a burst) rather than single bits in random positions. For example, a 1 millisecond impulse could corrupt 10 bits in sequence of a digital transmission at 9600 bits per second (bps). If the transmission rate was 2 Mbps then the number of bits corrupted could be 2000.

The parameter usually quoted as a measure of the quality of transmission systems is the bit error rate (BER), which is another name for the binary error probability used above. A BER of 10^{-5} means that on average one bit in every 100 000 will be corrupted. This is the figure usually given for the public switched telephone network. The quoted figure for BER will be a statistical mean and, obviously, subject to random variations.

Error control codes can be classified in several ways; for example, by the number of errors that can be detected or corrected in a block of binary symbols, or by subdivision into those whose purpose is the control of random errors and those whose purpose is the control of burst errors, or they can be put into one of two groups according to how they are generated. These latter two groups are known as *block codes* and *convolutional codes* respectively. Block codes contain a sub-set called cyclic codes, which have the property that any allowable code block forms another allowable code block when rotated. In general, the greater the power of the code the greater the redundancy needed for a given amount of data.

3.3.3.1 Block Codes

In simple terms, most block codes, whether error-detecting or correcting, operate in the same way. Data blocks (collections of data words, in turn collections of binary symbols) are mapped to code blocks in such a way that a change in any code block, whatever the cause, results in a set of symbols that are not allowed by the code. The fact that a non-allowed block is present must be detected by some kind of decoding mechanism and appropriate remedial action taken. If the code is such that it can only indicate that an error has occurred, the action to be taken depends on the application. In many communications systems, for example, a request for retransmission of that particular code block would be made. However, if the code allows for the correction of detected errors, then the affected bit(s) must be inverted before the data are passed on.

In the preceding paragraph the individual quantities of data and code were referred to as blocks. The most appropriate size of each data block to be translated into a code block is dependent on many factors. In real systems the optimum size can vary from as few as four bits to as many as several thousand bits. The factors that determine the optimum size of a block are very much application and code specific. However, in general terms it can be said that most codes are more efficient the longer the block length, that is, they need a smaller proportion of extra bits for a given error-detecting or correcting property. However, the longer the block length the more information is lost if the code breaks down. In addition, longer blocks usually take longer to encode and decode, so extra delays will be introduced into the system which may be unacceptable for real-time uses. So, for real systems the block length will be a compromise between these various factors.

Parity Codes

The simplest of all block codes involves the use of a single parity bit. A count of the number of 1s in the data block is made and an extra binary symbol is appended so that the number of 1s in the code block (data + parity bit) is even for even parity, or odd for odd parity. After transmission or storage the parity can be recalculated and tested against the original parity bit. If they are the same then the data are assumed to be correct. Clearly this method has many shortcomings, principally that it can only detect odd numbers of bits in error. If two or any other even number of bits are corrupted the parity bit will appear to be correct, as the number of 1s will still be an even (or odd) number. However, the hardware to generate parity is very simple for short data blocks so this method is in common use in semiconductor memory systems and in asynchronous communications where the basic unit of data is the 7-bit ASCII character, or the 8-bit EBCDIC character.

Product Codes

The concept of parity can easily be extended to cover a larger number of words of data, by storing them in an array of rows and columns. If n words, each of m bits, are to be stored then the array will have n rows and m columns, as shown below. (*Note*: the parity of the dotted sections is assumed to be even.)

```
                                                    ┌─ row parity bit
column parity bit ────► 0 1 1 0 1 1 0 1 ---- 1 0 0  ▼

data block              1 0 0 1 0 1 0 1 ---- 0 1 1
                        0 1 0 1 0 1 0 1 ---- 1 1 0
                        1 0 0 0 0 0 0 0 ---- 0 1 0
                        0 0 0 0 1 1 1 1 ---- 0 0 0
                        1 1 0 1 0 1 0 0 ---- 1 0 1
                        1 0 0 1 0 1 0 1 ---- 0 1 1
                        . . . . . . . . . . . . . . . . . . . .
                        . . . . . . . . . . . . . . . . . . . .
                        0 1 1 0 0 0 1 1 ---- 1 0 1
```

The parity can be calculated for each row as normal, and then for each column. The entire block of data and parity bits can then be transmitted. When a particular word is received the row parity can be recalculated. If this indicates an error then the column parity can be generated, once the entire block has been received, and can then be compared with that transmitted. Hence, the bit in error can be identified and corrected. The main advantage of this method of error control coding is the low redundancy needed for precise error correction; however it has many disadvantages, not least the time taken to calculate all the extra column parity bits. It is usually referred to as product coding and its main application has been in read-only memories, where the original parities need only be calculated once, rather than in communication systems. However, it can be used in very low data rate applications where the amount of redundancy is more important than the calculation time.

Hamming Codes

Most other block codes can be described as an extension of the basic parity concept. Parity is taken over various combinations of the bits in a longer data block and the generated check bits are used to form part of the code block. Such codes are usually denoted (n, k) where n is the total number of bits in the code and k is the number of data bits (that is, $(n - k)$ is the number of parity bits).

Classical Hamming Code

The simplest Hamming code is a single-error-detecting block code (with $d_{min} = 3$). It comprises the required number of information digits (for example, 5 for 32 symbols) interspersed with parity check digits. Each parity digit checks a unique selection of information digits. If a single error occurs in transmission, some of the parity checks will be found to fail. A 'checking number' is formed at the receiver, being a binary combination of the individual parity checks (writing '1' if it fails and '0' if it tallies). The selection of information digits checked by each parity digit is such that the checking number gives the actual position in the code word of the incorrect digit.

The positions in the final code word, and the selection of digits checked for each parity digit, are shown in Table 3.3.3.

Table 3.3.3

Parity digit	Position in code word	Digit positions checked
C_0	1	1, 3, 5, 7, 9, 11 ...
C_1	2	2, 3, 6, 7, 10, 11 ...
C_2	4	4, 5, 6, 7, 12, 13, 14, 15 ...
C_3	8	8, 9, 10, 11, 12, 13, 14, 15 ...

The checking number (in binary) is (C_3, C_2, C_1, C_0) with $C = 1$ for failure and $C = 0$ for a tally. The reason for the particular selection of digit positions will be illustrated by the example below.

EXAMPLE 3.3.3   A SINGLE-ERROR-CORRECTING/DOUBLE-ERROR-DETECTING HAMMING CODE FOR 16 SYMBOLS

We will ignore the double-error-detecting requirement initially. For 16 symbols we require 4 information digits ($2^4 = 16$). Since the check digits go in positions 1, 2, 4, 8 etc., we will have to put our information digits in positions 3, 5, 6 and 7, so three check digits are needed (in positions 1, 2 and 4) and the code is a (7, 3) type. Any suitable code can be used for the information digits, but a simple binary code has been used below in Table 3.3.4.

Table 3.3.4

		(C_0)	(C_1)		(C_2)				(P)
		1	2	3	4	5	6	7	8
(1)	A	0	0	0	0	0	0	0	0
	B	1	1	0	1	0	0	1	0
	C	0	1	0	1	0	1	0	1
	D	1	0	0	0	0	1	1	1
	.								
(10)	J	0	0	1	1	0	0	1	1
	.								
	O	0	0	0	0	1	1	0	1
(16)	P	1	1	1	1	1	1	1	1

The binary code is first written out in digit positions 3, 5, 6 and 7, leaving positions 1, 2 and 4 blank. C_0 is then placed in position 1, being an even-on-1s parity check on digit positions 1, 3, 5 and 7. The process is repeated for C_1 and C_2 using the combinations in Table 3.3.4 (ignore the digit shown in position 8 for the time being).

Suppose D (= 1 0 0 0 0 1 1) was transmitted and received as 1 0 1 0 0 1 1, that is, with the third digit in error. The parity checks are carried out, producing the binary checking number (C_2, C_1, C_0) = 0 1 1 = decimal 3. The third digit is in error and is therefore complemented and 'D' decoded. If no errors occur the checking number will be zero, but if two or more occur it will not provide meaningful results. The code can be made to detect double errors if an extra overall parity digit is added, and this has been placed in position 8 above, producing an (8, 4) code. Note that this is outside the Hamming process and must not be included in the error-correcting process.

The reason for the selection of digits checked by each check digit should now be clear. Since (C_2, C_1, C_0) is required to give the error position, C_0 must be 1 if the error is in positions 1, 3, 5, 7, 9 etc. since the binary equivalent of these numbers has a '1' in the least significant position, and similarly for C_1, C_2 etc. An (8, 4) Hamming code is used for the page headers and row-addresses in Teletext.

More Powerful Block Codes

The above treatment of the classical Hamming code can be used in simple cases, but a more general method is used for more complex applications. In mathematical notation, block codes are formed by multiplying the data or message block m by a generator matrix G (using modulo-2 arithmetic, that is, dropping the carries) to produce a code block C given by:

$$C = mG$$

If there are k symbols in the data block and n symbols in the code block, then the code will be denoted as (n, k) and C will be a $1 \times n$ vector, m a $1 \times k$ vector and G a $k \times n$ matrix. For example, for the data set in the simple (7, 4) Hamming code above, if m is the data sequence (1 0 0 1 – letter 'J' in the example) and G is the matrix

```
1 1 1 0 0 0 0
1 0 0 1 1 0 0
0 1 0 1 0 1 0
1 1 0 1 0 0 1
```

then the code word C will be 0 0 1 1 0 0 1, which is the same as obtained above; all the other codes can be generated correctly in the same way.

The ratio k/n is known as the code rate and is a measure of efficiency, in other words the amount of redundancy needed for a given error correction and detection capability. In general, linear block codes are more efficient the larger the block size but, as mentioned in previous sections, other factors about the system may limit the block size used in a particular application. A block code is said to be systematic if the data block sequence is an identifiable part of the code block, that is, the parity bits are appended to the data block to form the code block.

Decoding can be accomplished by multiplying the code block by a check matrix, H, derived from the generator matrix G in such a way that:

$$GH^T = 0$$

It can be shown that any correctly generated code block C will fulfil the requirement that:

$$CH^T = 0$$

This means that C is a valid code word if and only if it fulfils the requirement that when multiplied by the transpose of the check matrix H, a zero vector is the result. If the code block has been corrupted, the resulting block can be described by:

$$r = C + e$$

where e is the error pattern describing the corruption. The decoding process will yield a matrix S, known as the syndrome:

$$S = rH^T = CH^T + eH^T$$

As the first term has been defined as being zero, then $S = eH^T$.

In general, the pattern of non-zero bits in the syndrome can be used to indicate which bit(s) are in error. For a code block with no errors the syndrome will be all zeros. Using the same example again, the check matrix H becomes:

```
0 0 0 1 1 1 1
0 1 1 0 0 1 1
1 0 1 0 1 0 1
```

If the code word is received correctly (0 0 1 1 0 0 1), then performing the multiplication CH^T does indeed yield an all-zeros syndrome. However, if one of the bits has been corrupted, for example the code word becomes 1 0 1 1 0 0 1, then the syndrome will become 0 0 1, indicating that the first bit was the one in error. Other syndrome patterns will indicate the particular bit in error. The original data can then be recovered by performing the inverse multiplication with the generator matrix G.

The power of the code is determined by the relative values of k, the size of the data block, and n, the size of the code block. As described above, an important measure of the power is defined by the minimum Hamming distance, d_{min}, of the allowable set of code blocks. In the example used above, the Hamming distance for the code words is 3.

It was stated in section 3.3.3 that the error-correcting capability t of a code is given by

$$d_{min} \geq 2t + 1$$

and the error-detecting capability p by:

$$d_{min} \geq p + 1$$

It is possible to combine the detection and correction capabilities by using the form:

$$d_{min} = t + q + 1$$

where q is the number of additional bits that can be detected over and above those which can be corrected. Hence in the example code because d is 3, t is 1 and it is known as a single-error-correcting code. If more than one bit is in error then the code breaks down and will give incorrect results.

It is beyond the scope of this brief introduction to error control codes to show how the code block size for a given error-correction and detection capability is determined, that is, how k and n are related to d_{min}. The relationship is in terms of upper and lower bounds, relating the maximum and minimum values that k can take for a given n and a desired error-correction and detection capability. Various mathematicians have derived ways of determining the optimum position of these bounds, which come from consideration of the vector field space over which the particular code is defined.

For a single-error-correcting code, as used in the example, the relationship between n and k is straightforward:

$$2^{n-k} \geq n + 1$$

In the example, k was 4 and n was 7. If k is 16, n must be at least 21; if k is 32, n must be at least 38. Clearly the rate of the code (k/n) for a data block of 32 bits is greater than that for a data block of 16 bits (0.84 vs 0.76) and much greater than that for a data block of 4 bits (0.59). For Hamming-type codes the code block length n is always given by $2^m - 1$, where m is an integer greater than 3. The number of data bits k is given by $2^m - m - 1$.

Cyclic Codes

Cyclic codes are an important subdivision of block codes because they are relatively straightforward to encode and decode. They share two properties: firstly that any code word, when cyclically shifted to the right or the left, forms another code word, and secondly that the addition of one code word to another forms a valid code word. Some examples of the more important cyclic codes are described in the next few sections.

Bose–Chadhuri–Hocquenghem (BCH) Codes

BCH codes are a generalisation of Hamming codes that can be used for multiple-error correction. They allow for a wide variety of error-correcting capabilities, block lengths and code rates to be defined. Importantly, they can also be used with symbols that are not single binary digits. If a symbol is defined to be more than one binary digit then the preceding discussion about codes still holds good, except that the blocks are collections of symbols and not of single bits. A consequence of this is that BCH codes can be designed to correct bursts of bit errors. Let us say that a symbol of 5 bits is chosen and a code of double-error-detecting capability defined. This code will now be able to correct any two bursts of errors, five bits long or less, rather than merely two isolated bit errors. Clearly the block length of such a code, measured in bits, will be much greater than an equivalent one to correct isolated errors, but it illustrates the possibilities available to the system designer.

BCH codes are commonly used in telecommunications systems because their flexibility allows them to be tailored to suit the exact error-control needs of the transmission channel in use, whether burst or random. Some examples of block lengths for binary BCH codes are given in Table 3.3.5, where t is the error-correcting capability, and k and n are the data and code block lengths, as before.

Table 3.3.5   *Some examples of t-error-correcting binary BCH codes*

t	k	n	Code rate (k/n)
1	11	15	0.73
1	57	63	0.90
2	7	15	0.47
2	51	63	0.81
2	239	255	0.94
3	5	15	0.33
3	45	63	0.71
3	231	255	0.91
4	39	63	0.62
4	223	255	0.87
5	36	63	0.57
5	215	255	0.84
7	24	63	0.38
7	199	255	0.78
10	179	255	0.70
21	115	255	0.45
30	63	255	0.25

Notice that the code block length n, in Table 3.3.5, is given by $2^m - 1$, and that the desired error-correcting capability defines the number of information symbols that can be carried by that code block.

Reed–Solomon Codes

These are a sub-class of BCH codes that always use symbols larger than one binary digit. They are important because they achieve the largest possible code minimum distance for a given block length. As the symbols are no longer single binary digits, the minimum distance of the code is defined to be the minimum number of symbols in which the code sequences differ. The symbol error-correcting property t of a Reed–Solomon code is defined to be:

$$t = (d_{min} - 1)/2$$

and d_{min} is given by $n - k + 1$, where k is the number of symbols being encoded and n is the number of symbols in the encoded block.

Hence, in this case n, k and t can be related directly by:

$$n = k + 2t$$

Remember that n, k and t refer to symbols of more than one binary digit, so the error-correction capability will be in terms of groups or bursts of bits in error. The designer can trade off the amount of redundancy in the code with the error-correcting capability, encoding and decoding time, and other important system parameters.

Interleaving

A refinement that can be added to any block code to enhance its burst error-correction capability is that of interleaving. It is described here because it has a common practical application in conjunction with Reed–Solomon codes, in compact disc (CD) digital audio systems. The principle is that p consecutive code words, each made up of q symbols, are stored in an array of p rows and q columns. The symbols are then read out in columns and transmitted or stored, thus interleaving the symbols from any particular code word with all the other code words. The receiving system must first reassemble the array so that the rows (the original code words) can be read out to the normal block decoder. In this way a burst of errors of length p symbols can be tolerated as it can only affect two columns of the array at most and hence only one or two symbols per code word. The penalty is increased complexity and time to encode and decode. In the example of CD systems, the time penalty is irrelevant because it simply adds to the delay that all the bits go through and they still emerge from the decoder at the correct rate.

Cyclic Redundancy Check Codes

These codes, usually called CRC codes, are particularly important because of their widespread use in the data communications industry. They are er- ror-detecting codes only, so they provide no information whatsoever about which bits are in error. Hence, retransmission must be used as a method of error correction for systems employing CRC codes. Their power to detect both isolated and burst errors is very high and the necessary code rate is very low. The extra overhead of using an error-correction code to provide the detection capability of these uncomplicated codes would be prohibitive.

The basic principle behind them is simple. The data to be encoded are treated as a continuous string of binary symbols. This string is divided (modulo-2) by a known bit pattern and the remainder after division is ap- pended to the data that is being transmitted or stored. The quotient is dis- carded. When the data are received or read the division is recalculated, using the same divisor, and the newly generated remainder is compared with that accompanying the data string. Clearly, any difference between the two re- mainders indicates that the data have changed, in other words one or more errors have occurred.

In simple terms, the coding process can be set out as:

$$\frac{D}{G} = n + R$$

where D is the data stream, G is the generator polynomial (see below), n is the quotient of the division and is discarded, and R is the remainder to be appended to the data.

G is generally denoted as a polynomial in X of degree $m + 1$, where m is the desired length of the remainder, which determines the error-detecting power of the code. The notation requires that any bit position with a one in it is shown as a power of X, the index being the number of the bit position. When using this nomenclature, the bit pattern 10001000000100001 would be written as:

$$X^{16} + X^{12} + X^5 + 1$$

In fact this particular polynomial is an agreed international (ITU-T) standard bit pattern for use in data communications systems. It is 17 bits long and so will yield a 16-bit remainder. A Cyclic Redundancy Code (CRC), using this generator polynomial, has the power to detect the following:

All bursts of errors of length 16 bits or less
All odd numbers of bits in error
99.998 per cent of all error bursts greater than 16 bits

The last figure comes from the fact that the code will miss one in every 2^{16} error bursts greater than 16 bits in length, those that cause the division to give the same remainder as the original data. Remember that under this scheme the quotient is discarded, so only the remainder can be used for the checking process.

In some local area networks the CRC generator polynomial is 33 bits long, giving a 32-bit remainder. Bearing in mind the performance of the 16-bit CRC code shown above, this may seem excessive. However, the data rates used in local area networks tend to be so high that even very short electromagnetic pulses can cause corruption of a significant number of bits. For example, a 10 Mbps Ethernet subject to a burst of noise of 1 millisecond duration could have suffered corruption in ten thousand bits. A 32-bit error-checking code does not seem too much under such circumstances!

Another great advantage of CRC codes is the ease with which they can be encoded and decoded. All that is needed is a shift register and a few gates to perform the modulo-2 division, or it could easily be accomplished in software. In fact the vast majority of serial data interface circuits for microprocessors incorporate CRC generation and checking circuits. Some only allow the use of a fixed generator polynomial, usually the ITU-T-16 shown above, but others allow a choice of several other standards. Of course, the user has to ensure that both ends of the link are using the same generator polynomial or the receiver will make nonsense of the whole transmission.

Checksums

If the error-checking process has to be performed in software, the division needed by the CRC would take a prohibitively long time. In many data communications applications it is common to use a simple software checksum to perform error checking. In mathematical terms the data bytes that are to be encoded are added modulus 8, or 16, and the sum appended to the data. At the receiving end the addition process is re-performed and the new sum compared with that transmitted. The coding is extremely easy to perform, as modulo arithmetic in a binary system simply involves ignoring any overflow above the number of bits that are to be considered as the checksum. Hence the time taken to perform the checksum error-control algorithm is limited to the time the processor takes to add up the data bytes or words in the block. The error coverage of the checksum method is not as great as the CRC code, but the ease with which it can be performed in software makes it widely used.

Other Block Codes

Many mathematicians have devised particular codes or classes of code which have some parameter that has been optimised. Examples are the Golay code (for random errors) and Fire codes (for bursts of errors), both of which have found uses in digital communication systems. Any book on coding theory contains many examples of such codes; however those described above are by far the most commonly used.

3.3.3.2 Convolutional Codes

Convolutional codes are a class of error-correcting codes that differ from block codes in that the encoding process exhibits memory. If for every k input bits there are n output bits then the ratio k/n is called the code rate and it has the same significance as for block codes, that is, the amount of information carried per coded bit. However, the integers n and k are not sufficient to describe the lengths of the code and data blocks, as before. A third digit, K, known as the constraint length, is needed to characterise fully a convolution code. This is determined from the length of the memory element in the encoding circuit. The coded bits are a function, not only of the incoming data bits but also of the previous $K - 1$ data bits. As the process is continuous, no block lengths are defined. The block size that any system uses will be determined by other factors, not the encoding process. The code rate and the constraint length K give a measure of the error-correcting capability of a convolution code, but the relationship is not easy to define and is beyond the scope of this book. In general terms, a convolution code of a given rate will be more powerful than a block code of the same rate, that is, it will correct more errors for the same amount of redundancy. The trade-off is complexity in the decoding stage (see below).

Hagelburger's Code

This was one of the first convolution codes to be proposed (Hagelburger, 1959), and is illustrated in Figure 3.3.1.

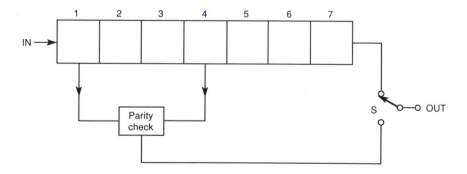

Figure 3.3.1 *Hagelburger's coding system*

The information digits are fed to a shift register and a parity check carried out between the first and fourth positions. The parity digit is then interleaved between successive information digits by means of the switch S. Table 3.3.6 shows the contents of the shift register and the corresponding output stream for several cycles for the data input 101011, starting with the shift register containing all zeros.

Table 3.3.6

Input bit	Shift register	Check digit	Output stream
	0 0 0 0 0 0 0	0	0 0
1	1 0 0 0 0 0 0	1	0 0 1 0
1	1 1 0 0 0 0 0	1	0 0 1 0 1 0
0	0 1 1 0 0 0 0	0	0 0 1 0 1 0 0 0
1	1 0 1 1 0 0 0	0	0 0 1 0 1 0 0 0 0 0
0	0 1 0 1 1 0 0	1	0 0 1 0 1 0 0 0 0 0 1 0
1	1 0 1 0 1 1 0	1	0 0 1 0 1 0 0 0 0 0 1 0 1 0
0	0 1 0 1 0 1 1	1	0 0 1 0 1 0 0 0 0 0 1 0 1 0 1 1
1	0 0 1 0 1 0 1	0	0 0 1 0 1 0 0 0 0 0 1 0 1 0 1 1 0 1

Because of the separation between the checked digit positions, the output stream is such that the separation between any two information digits covered by a parity check, or between any digit and its associated check digit, is always greater than six digit positions, so that bursts of six errors can be tolerated. Decoding is more complex, requiring the digits to be switched into data and check registers, with parity checks at appropriate points, and it can be shown that correct decoding is possible provided there are at least 19 correct digits between bursts of errors. The code rate k/n is one-half because there are two output bits for every input bit.

Encoding of General Convolution Codes

The structure of convolution codes is in general considerably more complex than the simple Hagelburger code described above, but they can be described diagrammatically using a trellis as shown in Figure 3.3.2.

A solid line represents a transition caused by a 0 in the input data and a dashed line represents a transition caused by a 1. Encoding is performed by following the path through the trellis determined by the incoming data and outputting the corresponding code word. In this example each data bit generates two code bits, hence the code rate is 1/2. If the hardware has to be flushed (that is, all trace of the previous block removed) after each arbitrary

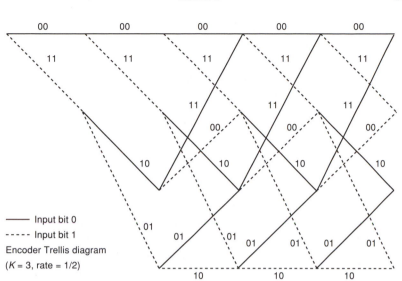

Figure 3.3.2 *Trellis for convolutional code*

length block then an extra *K* bits (all zeros) will be appended to the code block. Hence the true code rate will be worse than 1/2 for small block lengths but will approach 1/2 as the block length increases. It can be seen that after the first *K* stages the structure is repetitive (*K* = 3 in this case) and this is generally true.

As an example, if the data sequence 01001011 is passed through the trellis structure shown, the corresponding coded sequence becomes:

Data bit	0	1	0	0	1	0	1	1
Code bits	00	11	10	11	11	10	00	01

It can be seen that the code bits that are generated for each incoming 0 or 1 are dependent on what the previous data sequence was, as well as their own value.

Decoding of Convolutional Codes

Convolution codes can be decoded using a technique called *maximum likelihood*. This means that the incoming code stream is compared against every possible path through the trellis and the one that is closest to it defines the output data. A cost metric is used to measure the likelihood of a particular path having been taken. One possible cost could be the sum of the squares of the difference between the received values and the values of the outputs corresponding to that particular path. This cost metric, or any other that was chosen, would have to be calculated for every possible path. The path that has minimum cost is then chosen as the most likely and is used to provide the output from the decoder.

Of course, as the constraint length K increases, the complexity of the decoder must increase and the decoding operation rapidly becomes impractical. The requirement to store every possible path also places a major restriction on the decoder design. In practice an algorithm due to Viterbi is nearly always used to reduce the decoder complexity. In simple terms the Viterbi algorithm says that the minimum cost path to get to any point S_x in the trellis must have passed through some other point S_y. The path from the start to S_y must be the minimum cost path, otherwise the cost of the path to S_x could be lowered by changing the path from the origin to S_y. Hence minimum cost paths are built up from shorter minimum cost paths. This reduces, by a considerable margin, the need to store every possible path.

Another potential problem with convolution codes is that, in theory, there can be no output from the decoder until all the costs have been calculated and the minimum cost path chosen. Clearly this could be a major constraint on their usage in real-time applications. The Viterbi decoding scheme helps with this problem too, because the shorter minimum cost paths can be output before the final path is calculated.

The storage requirements and computational complexity of the decoding process restrict the constraint length of codes that can be used in real systems. Decoders for constraint lengths up to about 10 can be realised but these are powerful processors in their own right. A more practical limit to the constraint length K is about 6 or 7.

Trellis-Coded Modulation

One of the important applications of convolution coding is an attempt to combine the error-control coding of a digital signal with the modulation of that signal on to an analogue waveform for the purpose of transmission. This class of codes has become known as Trellis codes after the trellis-like structure described above for visualising the encoding and decoding operations of convolution codes. Trellis codes build upon multiple-phase shift keying (MPSK) or quadrature amplitude modulation (QAM) schemes to provide a coding gain, that is, a reduction in the tolerable signal-to-noise ratio for a given bit error rate (BER). Trellis codes are used by some modern modems to boost the achievable data rate over conventional telephone lines. In this case the coding gain is expressed as an improvement in the data rate for a fixed signal-to-noise ratio.

3.3.4 Examples for Solution

1. A simple language consists of four independent and equiprobable symbols A, B, C and D. The symbols are to be transmitted as a sequence of

marks and spaces through a binary symmetric channel of error probability 0.1. A, B, C and D are coded as 00, 01, 10 and 11 respectively; each '1' is transmitted as a sequence of three successive marks and each '0' as a sequence of three successive spaces. The receiver interprets each group of three received digits as '1' or '0' on a simple majority basis. Find the mean error rate for symbols.

2. (a) English text is to be transmitted via a binary symmetric channel that has a capacity of 50 binary digits per second and a binary error probability of 0.05. A non-redundant five-digit code is to be used for the symbols and punctuation (32 characters). Find the symbol error probability and comment on the quality of the received text.

   (b) A single parity digit is added to the above five-digit code and the system arranged such that a detected symbol error automatically leads to the immediate retransmission of that symbol until no error is detected. Find the symbol error probability and the effective rate of transmission of symbols, and comment on the quality of the received text.

3. (a) A single error-correcting and double-error-detecting binary Hamming code is required for 32 symbols, comprising letters and punctuation of English. Derive the first five code groups and find the efficiency of your code assuming that the symbols are equiprobable.

   (b) The above code is to be used in a binary symmetric channel of error probability 0.01. Find the probabilities of (i) a parity failure, (ii) two or more errors being detected, and (iii) a symbol error.

4. Verify that the syndrome generated by performing the matrix multiplication CH^T in example 3.3.3.1 is equal to 0.

5. A data communications link uses a CRC code with a generator polynomial of $X^3 + X^2 + 1$. Determine whether or not the following message sequences, received across this link, contain errors. For those sequences where there are no errors, give the valid data sequence:

   11001111 ; 11011101 ; 1011011001

4 Communication

4.1 Introduction

Communication and Information Theory are very closely related – indeed it is difficult to decide where one begins and the other ends. In this chapter we will review some of the important topics in Communication Theory in the light of the results of the previous chapters, and in preparation for their use in the next chapter.

We noted at the beginning of this book that information is rarely produced in a form that is suitable for immediate communication, and that it must be modified before transmission. We referred to this as coding, in a general sense, and to the reverse process at the receiver as decoding. It is usually called *modulation* in Communication Theory, where it involves impressing the information to be transmitted on to some form of 'carrier' waveform. For example, if we wish to transmit a speech waveform by radio, the speech cannot be broadcast directly because radio aerials transmit efficiently only at relatively high frequencies (when the aerial length is of the order of half a wavelength). The (low-frequency) speech waveform is therefore impressed on to a high-frequency sinusoidal carrier (which can be broadcast efficiently), which is said to be 'modulated' by the speech signal.

There are two basic types of modulation: analogue modulation and pulse modulation. In the former the carrier is a sinusoid and in the latter a stream of pulses, though in both cases the actual signal to be transmitted could be analogue or digital. Even in the latter case, however, the final transmission is often via a modulated sinusoid (for example, by switching a sinusoid between two or more levels), the pulse modulation being used only at an intermediate stage.

A communication system is characterised by a fixed bandwidth B and a specified signal-power-to-noise ratio, usually denoted SNR in communications applications. We saw in Chapter 1 that Shannon's Ideal Communication Theorem states that the capacity of such a channel is given by

$$C = B \log (1 + \text{SNR})$$

and that error-free communication is theoretically possible at any rate up to and including this value. Bandwidth and signal-to-noise ratio may be exchanged at a fixed capacity, and modulation schemes that tend to spread the signal energy over a wide bandwidth were seen to be particularly attractive in terms of the low power required for transmission. We will compare the various modulation methods on this basis and also with the limits set by the Ideal Theorem.

128

4.2 Analogue Modulation

A sinusoidal waveform of amplitude a_c, angular frequency ω_c and phase ϕ can be described by the equation $v_c(t) = a_c \cos(\omega_c t + \phi)$. We have seen that a pure sine wave does not carry information, since it is totally predictable. However, we can impress information on to it by modifying a_c, ω_c or ϕ in sympathy with the signal to be transmitted. Methods for modifying a_c are known as *envelope modulation* or as *amplitude modulation*; those in which the instantaneous phase $(\omega_c t + \phi)$ of the sine wave is modified are known as *angle modulation*.

4.2.1 Amplitude Modulation

When a carrier $v_c = a_c \cos \omega_c t$ is amplitude modulated by a signal $v_s = a_s \cos \omega_s t$, the resulting waveform is given by

$$v_m = a_c(1 + m \cos \omega_s t) \cos \omega_c t \qquad (4.2.1)$$

where m is known as the *modulation index*, or *depth of modulation* (if expressed as a percentage).

$$m = \frac{\text{amplitude of the modulating wave}}{\text{amplitude of the carrier wave}} = \frac{a_s}{a_c}$$

The AM waveforms are sketched in Figure 4.2.1(a). It can be seen that the signal v_s determines the envelope of the modulated waveform; v_m can range in amplitude from $a_c(1 + m)$ to $a_c(1 - m)$ and clearly cannot be allowed to become negative (or the correct waveform would not be recovered from the envelope), so we have the restriction that $|m| \le 1$.

Equation 4.2.1 can be rewritten in the form

$$v_m = a_c \cos \omega_c t + \frac{ma_c}{2} \cos(\omega_c + \omega_s)t + \frac{ma_c}{2} \cos(\omega_c - \omega_s)t \quad (4.2.2)$$

showing that it consists of three terms, the carrier and two side frequencies, which constitute the spectrum of the modulated waveform for the case of modulation by a sinusoidal signal. This form of amplitude modulation is known as Double–Sideband, Large Carrier AM or DSB-LC.

The form of the spectrum for a more general signal $f(t)$, such as a speech waveform, is easily seen by convolution. Equation 4.2.1 becomes

$$v_m = a_c(1 + m f(t)) \cos \omega_c t = a_c m f(t) \cos \omega_c t + a_c \cos \omega_c t$$

The first term is a product between $f(t)$ and $\cos \omega_c t$ and the spectrum is therefore the convolution between the spectrum of $f(t)$ and impulses at $\pm f_c$; the second term (the unchanged carrier) simply produces impulses at $\pm f_c$. This is illustrated in Figure 4.2.1(b). The spectrum $F(f)$ of the speech waveform

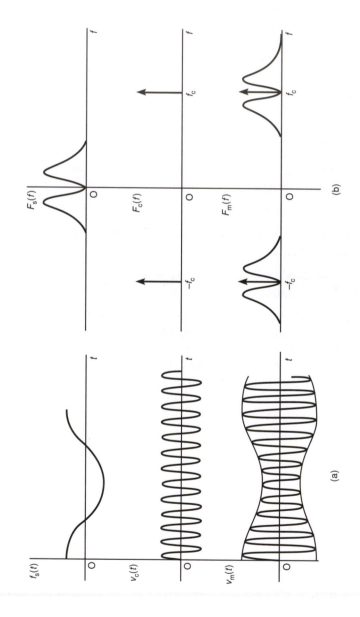

Figure 4.2.1 *Amplitude modulation (a) and its spectrum (b)*

$f(t)$ is simply reproduced (scaled by m) centred on the carrier frequency, together with the unchanged carrier. The two sections of the spectrum $F(f)$, above and below the carrier frequency f_c, are known as sidebands. The effect of AM is thus simply to shift the spectrum by an amount equal to the carrier frequency, placing it at a suitable frequency for transmission. The appearance of the two sidebands is a direct consequence (and an instructive illustration) of Fourier Theory; if one insists on representing spectra in terms of only positive frequencies, one finds nevertheless that a second (mirror-imaged) sideband miraculously appears, whereas using Fourier representation the complete two-sided spectrum is simply shifted up to f_c (and down to $-f_c$).

From Figure 4.2.1(b) it can be seen that all the information in the modulating signal is present in each of the sidebands, and that no information is present in the carrier term. Consequently it is possible to suppress (that is, not transmit) the carrier and one of the sidebands, and still get all the information across. If both sidebands are transmitted, without the carrier, then the method is known as Double-Sideband, Suppressed Carrier AM or DSB-SC. If the carrier and one of the sidebands are suppressed, then the method is known as Single Sideband AM or SSB.

The saving in transmitter power that is achieved by using DSB-SC or SSB can be deduced from equation 4.2.2 by considering the following relationships:

$$\frac{\text{carrier power}}{\text{total power}} = \frac{a_c^2}{a_c^2(1 + \frac{1}{2}m^2)} = \frac{2}{2 + m^2}$$

$$\frac{\text{power in each sideband}}{\text{total power}} = \frac{\frac{1}{4}ma_c^2}{a_c^2(1 + \frac{1}{2}m^2)} = \frac{\frac{1}{2}m^2}{2 + m^2}$$

If $m = 0.5$ (a typical value for many applications), then the power in the unmodified carrier is 89 per cent of the total transmitted power. The power in each of the sidebands is about 5.5 per cent of the total power. Hence for DSB-LC only 5.5 per cent of the transmitted power is needed to carry the information. However, because only part of the envelope of the AM wave is present in both DSB-SC and SSB, recovery of the information at the receiver is very much more difficult.

4.2.1.1 Generation of Amplitude Modulation

Modulation and demodulation of AM are non-linear processes and both can be performed by multiplication. We saw above that a characteristic of a linear system is that no frequencies are produced at the output that were not present at the input (only amplitude and phase changes occur), whereas the spectrum is actually shifted in frequency in AM.

AM can be produced by multiplying a carrier by the signal:

$$\cos \omega_c t \cdot \cos \omega_m t = \tfrac{1}{2} \cos(\omega_c + \omega_m)t + \tfrac{1}{2} \cos(\omega_c - \omega_m)t$$

which is the waveform of DSB-SC AM. In order to get full AM (DSB-LC) the DSB-SC waveform must be suitably scaled and added to the original carrier (see Figure 4.2.1b).

Single-Sideband AM can be produced by filtering out the unwanted sideband from a DSB-SC wave. This is only practical if the lower cut-off frequency of the signal bandwidth is not too low, otherwise the two sidebands are very close together and filters with impossibly sharp roll-off characteristics would be required. SSB can be generated directly by adding the outputs of two DSB-SC modulators, with one set of inputs shifted by 90 degrees, as shown in Figure 4.2.2.

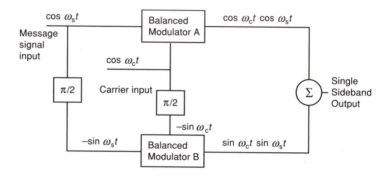

Figure 4.2.2   *Single-Sideband Modulator*

The output of DSB-SC modulator A is given by

$$\cos \omega_c t \cdot \cos \omega_m t = \tfrac{1}{2} \cos(\omega_c + \omega_m)t + \tfrac{1}{2} \cos(\omega_c - \omega_m)t$$

and the output of DSB-SC modulator B by

$$\cos(\omega_c t + \pi/2) \cdot \cos(\omega_m t + \pi/2) = \tfrac{1}{2} \cos(\omega_c + \omega_m + \pi)t + \tfrac{1}{2} \cos(\omega_c - \omega_m)t$$

Noting that $\cos (A + \pi) = -\cos A$, it can be seen that adding these two together produces a single-sideband signal: $v = \cos(\omega_c - \omega_m)t$. If the upper sideband is required then one of the phase shifts must be replaced by a $\pi/2$ lag instead of a lead.

4.2.1.2   Demodulation of AM

If all the envelope of the AM wave is present at the receiver, then a simple method of demodulation using a diode can be employed; the diode is actually switched on and off by the carrier (a non-linear process equivalent to multiplication) and a smoothing filter recovers the signal from the resulting

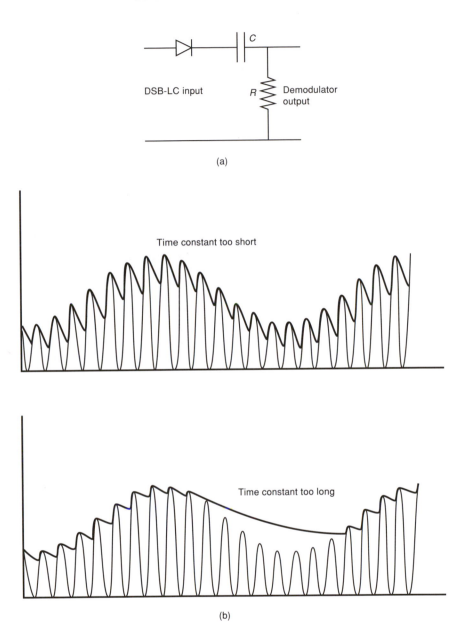

Figure 4.2.3 *Envelope detection: (a) demodulator, (b) effect of time constant*

half-sine waves, see Figure 4.2.3. For obvious reasons, this is known as *envelope detection*.

If one or more of the original components of the envelope are not present at the receiver, then envelope detection cannot be used. For DSB-SC and SSB, demodulation involves a further multiplication by the carrier; in Figure 4.2.3 the original spectrum can be recovered by convolving the spectrum of the modulated waveform by impulses at the carrier frequency, that is, by a multiplication by the carrier in the time domain. This technique is known as *coherent detection*.

A practical difficulty that arises with coherent detection is the need to ensure that the frequency and phase of the carrier wave used by the demodulator are exactly the same as those used by the original modulator. If the carrier at the receiver (usually known as the *local carrier*) has the form $v_{lc} = \cos[(\omega_c + \Delta\omega)t + \phi]$, implying a frequency and phase difference from the original carrier, then the result of multiplying it with the incoming AM wave (which, for DSB-SC, will be of the form $v_{am} = \cos \omega_c t.\cos \omega_m t$) will produce

$$v = K \cos \omega_c t \cos[(\omega_c + \Delta\omega)t + \phi]\cos\omega_m t$$

where K is a constant determined by the amplitude of the received signal after amplification.

If the first two terms are combined (using the standard trigonometric relationship for $\cos A \cos B$) then it can be seen that this wave has components at frequency $2\omega_c$ which can be removed by a low-pass filter, leaving

$$v = k \cos \omega_m t.\cos[\Delta\omega t + \phi]$$

Again, the amplitude k is controlled by the precise form of the receiver electronics. This signal is the desired information ($\cos \omega_m t$) multiplied by a term representing the frequency and phase errors of the local carrier. If $\Delta\omega = 0$ and $\phi = 0$, then the coherent detector has recovered the original signal. However, if either error is non-zero (or as is likely in a real system, is varying with time), the received information will have a time-varying distortion superimposed upon it. Practical demodulators use a replica of the original carrier derived from the DSB-SC waveform, to overcome this limitation. One circuit arrangement which is commonly employed, using two coherent detectors, is known as a *Costas Loop*.

Because of the difficulty (and hence expense) of mass-producing practical coherent detectors (as well as the enormous inertia preventing a change from the current system), all AM radio broadcasting uses DSB-LC, despite the tremendous waste of transmitter power and bandwidth. This means that cheap envelope detectors can be used in radio receivers. However, both DSB-SC and SSB are used in point-to-point radio links. The extra expense in the receiver is more than offset by the saving of transmitter power.

4.2.1.3   Vestigial Sideband Modulation

Television broadcasts use a compromise between the three main types of AM, known as *Vestigial Sideband* modulation (VSB). This is illustrated in

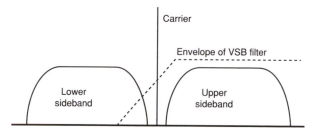

Figure 4.2.4 *Vestigial Sideband Modulation*

Figure 4.2.4. One of the AM sidebands is transmitted without attenuation, but the carrier term is severely attenuated before transmission and only part of the other sideband is sent. This has the advantages of power and bandwidth saving, while transmitting enough of the original carrier to allow the receiver to phase and frequency lock a local oscillator. Hence there is no need for a complex coherent detector, reducing the cost of television sets.

4.2.1.4   Effect of Noise on AM Systems

In real systems, where information has to be recovered from an amplitude-modulated signal, the effects of demodulation on the signal-to-noise ratio must be considered. For DSB-LC signals, most of the power in the modulated wave is in the carrier, which is removed by the demodulator, so it is not surprising that the SNR at the output of an AM demodulator is worse than the SNR at the input. The relationship can be shown to be:

$$\text{SNR}_{\text{out}} = \frac{2m^2}{2 + m^2} \, \text{SNR}_{\text{in}}$$

For DSB-SC, all of the incoming signal is carrying information but the bandwidth at the output of the demodulator is half that at the input, so a 3 dB increase in SNR can be achieved by the demodulation process. It should be noted that for equal transmitted power, a DSB-SC system will have a far superior SNR at the receiver, when compared with DSB-LC. This is because of the enormous waste of power caused by the transmission of the carrier wave.

For SSB, the demodulation process will have no effect on the SNR, as all that is performed is a frequency shift in the signal – there is no reduction in bandwidth. From these statements, it could be argued (and often is in textbooks) that DSB-SC AM has a superior performance in the presence of noise than SSB AM. However, a Single-Sideband system only requires half the transmitted power of a Double-Sideband system, for the same range. Hence, for equivalent transmitter powers, the SNR at the output of the demodulator of a DSB-SC system is the same as that of a SSB system.

4.2.2   Frequency Modulation

There are two forms of angle modulation, known as phase modulation and frequency modulation. Phase modulation (PM) modifies the phase angle ϕ and frequency modulation (FM) the instantaneous angular frequency ω_c. Phase modulation was rarely used for analogue communications, because of its variable bandwidth requirements. However, it is now very common in data communications systems over telephone channels, because it can give a more efficient use of the limited bandwidth.

Figure 4.2.5 illustrates frequency modulation. A sinusoidal signal $v_s = a_s \cos \omega_s t$ modulates the carrier $v_c = a_c \cos \omega_c t$ by varying the instantaneous frequency f_i. The modulated waveform v_m is given by $v_m = a_c \cos \theta_i$, where θ_i is the instantaneous angle. The instantaneous angular frequency is therefore given by

$$\omega_i = \frac{d\theta_i}{dt} = 2\pi f_i$$

If we define $f_i = f_c + k\, a_s \cos \omega_s t$, where $ka_s = \Delta\omega$ is known as the peak angular frequency deviation, then

$$\theta_i = \int (2\pi f_c + 2\pi \Delta f_c \cos \omega_s t) dt = \omega_c t + \frac{\Delta f_c}{f_s} \sin \omega_s t + c$$

If $\theta_i = 0$ when $t = 0$, then $c = 0$.

Defining β, the modulation index, as

$$\beta = \frac{\Delta f_c}{f_s} = \frac{\text{peak frequency deviation of the carrier wave}}{\text{frequency of the modulating wave}}$$

we obtain the standard equation for an FM wave:

$$v_m = a_c \cos(\omega_c t + \beta \sin \omega_s t) \qquad (4.2.3)$$

It can easily be verified from this equation that the instantaneous frequency is $f_c + \Delta f_c \cos \omega_s t$ where the peak deviation $\Delta f_c = \beta f_s$, as defined above. The instantaneous frequency thus varies sinusoidally (at frequency f_s) about the carrier frequency f_c between $f_c - \Delta f_c$ and $f_c + \Delta f_c$.

Note that there is no need for a restriction on β (as there was on m for AM) since f_c is usually several orders of magnitude higher than f_s. FM broadcasting is usually restricted to a maximum deviation of the carrier frequency of ± 75 MHz. If the maximum frequency present in the information signal is taken to be 15 kHz, this gives an equivalent value for β of 5.

Unfortunately it is not possible to obtain the spectrum of FM by convolution! This is because we require a cosine of a sine, as in equation 4.2.3, or more generally a cosine of some function $f(t)$. Such requirements always produce rather unpleasant mathematical entities known as Bessel functions,

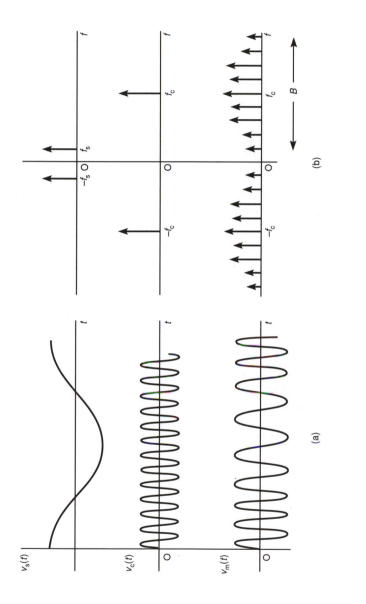

Figure 4.2.5 *Frequency modulation (a) and its spectrum (b)*

which are tabulated functions of an oscillatory form (see Appendix 4). It can be shown that the modulated waveform v_m can be written as

$$v_\mathrm{m} = \sum_{n=-\infty}^{\infty} a_c J_n(\beta)\cos(\omega_c + n\omega_s)t$$

where $J_n(\beta)$ are Bessel functions of order n. The form of the spectrum is sketched in Figure 4.2.5(b) and consists of pairs of side-frequencies at $f_c \pm nf_s$.

The magnitude of the side-frequencies depends on $J_n(\beta)$, the carrier not necessarily being the largest, but $J_n(\beta)$ becomes small for $n > \beta + 1$. The total power in the waveform is constant, since the amplitude is unchanged, but the distribution of power with frequency depends strongly on β. The channel bandwidth B_c is given approximately by *Carson's rule*:

$$B_c = 2(\beta + 1)\,B_s$$

For small β (< 0.2), $B_c \approx 2B_s$, and the modulation scheme is known as *Narrow-band FM (NBFM)*. The modulated wave is very similar to DSB-LC AM in that it consists of a carrier term and two side-frequencies. However, the phase and amplitude relationships between the various components are very different from DSB-LC. For large β (> 5), $B_c \approx 2\beta B_s = 2\Delta f$, which is much greater than the signal bandwidth. This is known as *Wide-band FM*. Hence Wide-band FM has the desirable property, deduced from the ideal communication theorem, that it automatically spreads the signal energy over a wide bandwidth.

4.2.2.1   Generation of FM

FM can be produced by a voltage-controlled oscillator, whose instantaneous frequency depends on the input signal amplitude. However, for broadcast purposes, where a stable carrier (or centre) frequency is vital, a VCO cannot be used. A method using a so-called *Armstrong modulator* is commonly used, shown in Figure 4.2.6.

A Narrow-band FM signal is produced, by integrating the information signal and then phase-modulating a carrier. This is then frequency multiplied, which increases the value of β and the carrier frequency. If necessary,

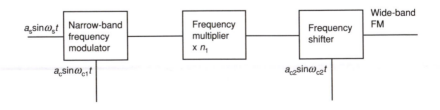

Figure 4.2.6   *Armstrong modulator*

the carrier frequency can be reduced (shifted) by a balanced modulator stage. It is possible to have multiple-frequency multiplication and frequency shift stages in order to achieve whatever final value of β and carrier frequency are wanted.

For example, to generate a Wide-band FM output with 100 MHz carrier and $\beta = 5$, from an NBFM wave with 8 MHz carrier and $\beta = 0.2$, the required stages could be:

(i) frequency multiply by 10, giving $f_c = 80$ MHz, $\beta = 2$
(ii) frequency shift by mixing with 40 MHz, giving $f_c = 40$ MHz, $\beta = 2$
(iii) frequency multiply by 2.5, giving the desired output: $f_c = 100$ MHz, $\beta = 5$.

4.2.2.2 Demodulation of FM

FM can be demodulated by a *frequency discriminator* circuit, which is a differentiator followed by an envelope detector. If we refer back to the defining equation for an FM wave, $v_m = a_c \cos \theta_i$, and differentiate, we get

$$v_m = a_c \, \dot{\theta_i} \sin \theta_i$$

As $\dot{\theta_i} = \omega_c + K v_s$ (where K is a constant), the waveform is an FM wave with an amplitude proportional to the original modulating signal. The variations in frequency of the carrier wave will not affect the operation of the envelope detector, whose output will be the desired information signal.

Another method of demodulating FM is via a phase-locked loop, in which a voltage-controlled oscillator (nominally at the carrier frequency) in a feedback loop follows the instantaneous frequency, the necessary controlling voltage being proportional to the signal. In either case the input waveform is usually amplitude-limited to remove any unwanted amplitude variations.

4.2.2.3 Effect of Noise on FM Systems

The noise power at the output of an FM demodulator rises rapidly with frequency, as shown in Figure 4.2.7. However, since the signal bandwidth is much less than the channel bandwidth, most of the noise can be rejected by a filter. The demodulator thus fully extracts the signal from the FM waveform but is able to reject most of the noise, and produces a very favourable signal-to-noise ratio, compared with AM, for a given transmitted power. It can be shown that the signal-power-to-noise ratio $(SNR)_o$ at the output is related to that on the channel $(SNR)_c$ by

$$(SNR)_o = 3\beta^2(\beta + 1)(SNR)_c$$

Intuitively one can see that in an FM system the information is in terms of frequency, which is less easily affected by noise than the amplitude in an

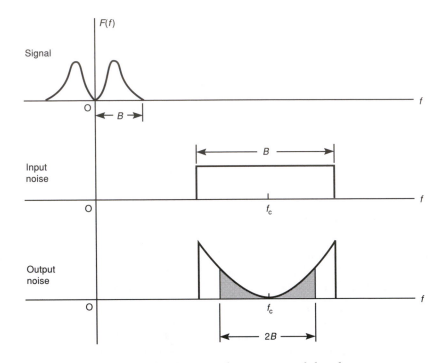

Figure 4.2.7 *Rejection of noise in a frequency-modulated system*

AM system; similarly, a sideband is repeated as many separate sidebands in FM, but only as two in AM. The improvement in the SNR of FM systems, due to the demodulation process, is known as *quieting*.

For a fair comparison of the performance in the presence of noise of FM and AM systems we must constrain the channel bandwidths to be equal. Using Carson's rule for FM, and the results given above for DSB-LC AM, it can be seen that

$$SNR_{out(FM)} = \frac{3(2 + m^2)}{2m^2} \beta^2 \; SNR_{out(AM)}$$

For a radio broadcast, with $m = 0.3$ and $\beta = 5$, this gives $SNR_{out(FM)} = 870 SNR_{out(AM)}$. In other words, for an equivalent SNR at the detector output, an AM system needs about 29 dB more transmitter power than an FM system. The FM system spreads the signal power over a wide bandwidth $(2(\beta + 1)B_s)$ but is transmitted at a much lower amplitude level than AM, making the system much more efficient as predicted by the ideal communication theorem.

A further improvement in the noise performance of FM systems can be obtained by a technique known as pre-emphasis, in which the higher-frequency components of the signal are boosted before transmission, with the reverse operation (de-emphasis) applied at the receiver. The higher-frequency noise components are thus further reduced, so $(SNR)_o$ is improved.

A threshold effect occurs in all practical systems that exchange bandwidth for signal-power-to-noise ratio. In the case of FM, if we try to make B_c too large this makes $(SNR)_c$ so low that noise causes unwanted zero crossings to occur. This causes a dramatic fall in $(SNR)_o$, so in practice a minimum $(SNR)_c$ of about 5 has to be maintained. The threshold and quieting effects are illustrated in Figure 4.2.8.

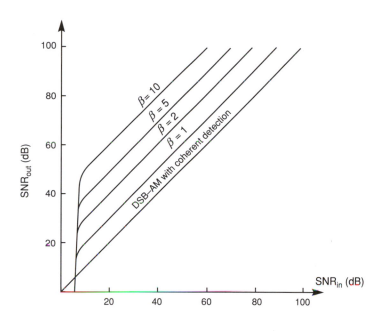

Figure 4.2.8   *SNR performance of Wide-band FM*

4.3   Pulse Modulation

In many practical situations the information to be transmitted is in the form of pulses. For example, if several signals are to be transmitted via a single channel they may be sampled in turn and successive samples transmitted, producing a stream of pulses of different height (this process is known as Time Division Multiplexing (TDM); a similar process known as Frequency Division Multiplexing (FDM) can be used in the frequency domain). There are a number of methods by which the information in the pulses may be modified before transmission, to improve the tolerance of noise or to reduce the power required, and these are collectively known as pulse modulation systems.

4.3.1 PAM, PWM and PPM

The principal methods are illustrated in Figure 4.3.1, in which the signal is a triangle wave. In pulse amplitude modulation (PAM) the signal is simply sampled directly at a suitable rate, that is, greater than $2B/s$ where B is the signal bandwidth, producing a stream of pulses of height proportional to the signal. The pulse width is usually made small compared with the time between samples (so that samples from other signals can be interleaved if required). We have seen that the spectrum of the sampled waveform consists of repetitions of the signal spectrum spaced at the sampling frequency $1/T_s$, and the bandwidth required for PAM is usually taken to be $1/T_s$. PAM is similar to AM in that the bandwidth is essentially the same as the signal bandwidth and it is therefore inefficient in terms of power required for a given output-power-to-noise-ratio.

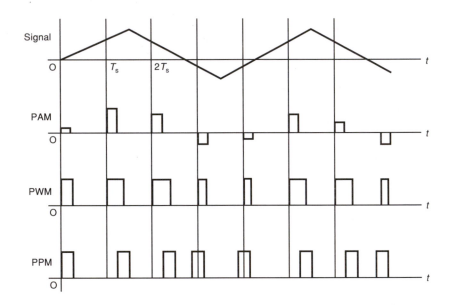

Figure 4.3.1 *Pulse modulation systems*

The information in a pulse-width modulation system (PWM, sometimes called PDM for pulse duration modulation) is in terms of the width of the pulses, a standard width for zero signal being increased or decreased dependent on signal polarity. Similarly, in pulse-position modulation (PPM) the information is in the position of the pulse with reference to a standard position corresponding to zero signal. In both these cases the shape of the pulse is important, as the leading or trailing edges have to be located accurately. A trapezoidal pulse with rise time τ_r is often used and the required channel bandwidth taken as $1/(2\tau_r)$. An infinite bandwidth is needed for perfect transmission, so any practical value is a compromise. PWM and PPM are a little like FM in that the channel bandwidth is greater than the signal bandwidth, and are therefore more efficient than PAM. However, they have been largely replaced by pulse code modulation (PCM) in recent years, though PWM in particular is now widely used in control systems and also in modern analogue/digital converters.

4.3.2 Pulse Code Modulation

In PCM the signal is first sampled as in PAM and the samples then quantised to the nearest of a preset number of levels. The quantised values are then converted to binary numbers and a stream of corresponding binary digits representing the successive samples transmitted. The information is neither in the pulse height nor its position, but in the actual sequence of 0s and 1s. Figure 4.3.2 illustrates the process for a simple system with eight levels.

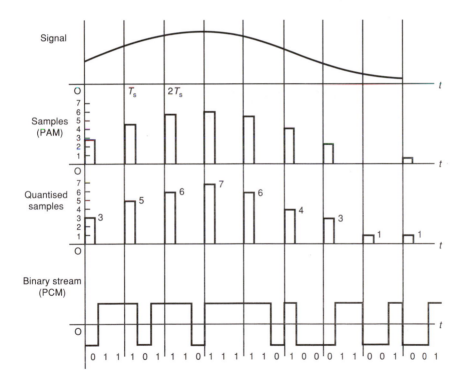

Figure 4.3.2 *Pulse code modulation*

In general, if we have a signal bandwidth B_s then we must sample at more than $2B_s$/s. In a binary system the total number of levels M is given by $M = 2^m$, where m is the number of pulses required to represent each sample (in Figure 4.3.2, $M = 8$ and $m = 3$). Using the general rule that a channel of bandwidth B_c can accept $2B_c$ binary pulses/s, the channel bandwidth required is $2B_s \times m \times \frac{1}{2} = mB_s$ (sampling at exactly $2B_s$). It is therefore much greater than the signal bandwidth, so that the power required for transmission will be low, as seen from the Ideal Communication Theorem in Chapter 1.

EXAMPLE 4.3.1

Compare the channel bandwidths required for PAM and PCM for a 3 kHz telephone channel sampled at 8000/s; the PCM system has 128 levels and one extra synchronising bit.

The PAM system simply requires $1/T_s = 8$ kHz. The PCM system has $M = 128$ levels plus the additional synchronising bit, giving $m = 7 + 1 = 8$. The bandwidth required is therefore half the number of binary pulses/s, which is $\frac{1}{2} \times 8000 \times 8 \approx 32$ kHz.

The particular advantages of PCM are that it is very tolerant of noise and very efficient in terms of power required for transmission. It is only necessary to determine at the receiver whether the received digit is a '0' or a '1', and provided that binary errors do not occur the system is noiseless! In fact, binary errors must not be allowed to occur, since their effect would clearly be disastrous, so the digits must be transmitted at a sufficiently high signal-to-noise ratio to make the probability of an error negligibly low. This is another example of the threshold effect mentioned above; if one reduces the power too much the system rapidly becomes useless. Fortunately the power-to-noise ratio required for an acceptable error rate is surprisingly low. Using the method described earlier for finding the error rate of a binary signal perturbed by Gaussian noise, the critical value is found to be about 20 (corresponding to a root mean square signal-to-noise ratio of 4.6) for an error probability of 10^{-5}; increasing the signal above this value produces a rapid decrease in errors.

Although PCM is noise free in the sense described above, things are rarely ideal in theory, let alone in practice. When PCM is decoded it exactly re-produces the quantised values, not the original samples, and the difference between the two is known as quantising noise, discussed in section 2.5.5. It was shown there that with a signal range of $s/2$ to $- s/2$ and with an m-bit code giving M ($= 2^m$) equally spaced quantising levels (with spacing s/M) the output-power-to-noise ratio is given by $(SNR)_o \approx M^2$.

In practice, many PCM systems use non-equally spaced quantisation steps, to allow for a greater dynamic range in amplitude. Smaller amplitude signals occupy a disproportionally large part of the available code space. This technique

is known as *companding*. In Europe, a method known as A-Law companding
is specified for use by the telecommunications operators. This is illustrated
in Figure 4.3.3, and is described by the following equations:

$$\text{For } 0 \leq V_{in} \leq \frac{1}{A} \qquad V_{out} = \frac{AV_{in}}{1 + \ln A}$$

$$\text{For } \frac{1}{A} \leq V_{in} \leq 1 \qquad V_{out} = \frac{1 + \ln(AV_{in})}{1 + \ln A}$$

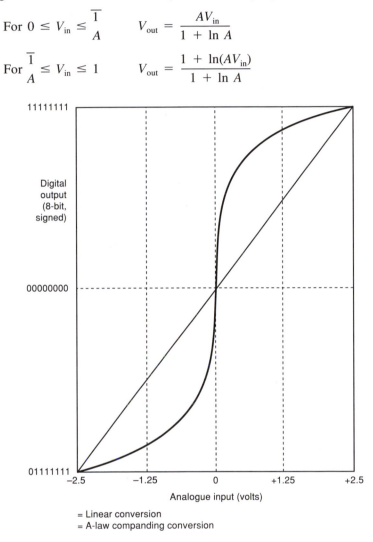

= Linear conversion
= A-law companding conversion

Figure 4.3.3 *A-Law companding*

The standard value for A is specified as 87. In the USA a different companding
scheme, known as μ-law, is used. The difference in achieved signal-to-noise
ratio between A-law and μ-law is small.

A further advantage of PCM is that binary pulses can be processed much
more easily than in systems where pulse height or position is important. In

many communication systems it is necessary to place 'repeaters' at suitable intervals to reshape the signal before unacceptable attenuation or distortion occurs. In PCM one simply has to produce clear ones and zeros again, so the process is very straightforward. Moreover, no information is lost in the process (unlike PAM etc.) since the original data are exactly restored (assuming no binary errors again). This, of course, is the desirable characteristic of digital systems, and the same advantages apply for example to digital audio discs as opposed to conventional analogue discs.

EXAMPLE 4.3.2

Compare the efficiency of PCM with that of PAM and with an ideal communication system.

Consider a 3 kHz telephone signal sampled at 8000/s. We will assume a PCM system with 256 levels ($m = 8$) with channel power-to-noise ratio 20 (giving negligible binary error). The channel bandwidth will be $\frac{1}{2} \times 8 \times 8000 = 32$ kHz compared with the PAM bandwidth of 8 kHz. If the transmitted powers are the same, the PAM system will have a channel power-to-noise ratio of $20 \times (32/8) = 80$ and the same value at the output. The PCM system will have output-power-to-noise of $M^2 = (256)^2$ or about 65 000. A similar ratio applies to the transmitted powers required for the same output-power-to-noise.

Considering an ideal system, we have $C = B \log (1 + S/N)$. The capacity of the PCM system is $8 \times 8000 = 64\,000$ bits/s, being the number of binary digits transmitted per second. For the same capacity and bandwidth an ideal system would have a channel power-to-noise ratio of 3, requiring only about 1/7 (20/3) of the power of PCM.

4.3.2.1   Differential PCM

If the sampling frequency of a PCM system is much greater than the rate of change of the analogue signal that is being sampled, it is possible to predict the next quantised value from a knowledge of the previous values. Of course, there will often be an error between the real and the predicted values. However, the value of the errors will be considerably smaller than the range of the original signal. Differential PCM (DPCM) makes use of this by coding a prediction error rather than the real signal. As the error is smaller it will require fewer quantisation levels, and hence fewer bits to transmit. A further step can be taken by making the quantisation of the error signal adaptive – if at any time the average value of the error signal is small, the quantisation step size will be decreased so that the sampled values are closer to the real values. As the average value of the error signal increases, so do the quantisation step sizes, keeping the number of bits to be transmitted down. This technique, known as *Adaptive Differential PCM* (ADPCM), can reduce the number of bits per second needed to transmit speech over a tele-phone channel by a factor of about 2.

4.3.3   Delta-modulation

A completely different approach to the analogue-to-digital conversion process, known as *delta-modulation*, can be used to produce a single-bit representation of the information waveform. A delta-modulator compares the current sampled level with the previous sampled level. If it is higher, then the modulator emits a positive level (or pulse). If it is lower, it emits a negative or zero level (or pulse). Clearly, the sampling rate has to be high enough so that the output of the modulator can follow a rapidly changing waveform. If it is not then so-called *slope overload*, which is a form of distortion, occurs. Figure 4.3.4 shows the waveforms associated with delta-modulation.

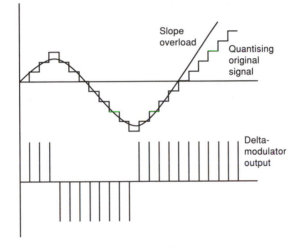

Figure 4.3.4   *Delta-modulation*

Demodulation can be achieved by a simple low-pass filter. There are many variations in implementation for delta-modulators, including adaptive ver-sions which avoid slope-overload under most conditions. However, these remove the main virtue of delta-modulation which is the simplicity of the circuits needed to implement it. The main applications of delta-modulation are in tape-recording systems and in military secure speech systems.

4.4 Binary Communication Systems

PCM is an effective technique for carrying information, but in pure binary form it does not give the receiver all that it needs to know in order to recover that information. For example, in order for the receiver to be able to distinguish individual bits, it is important that it should be able to recover the clock rate from the incoming waveform, to avoid the need for separate transmission of clocking information. This would be impossible with the data in pure binary format (1 = high, 0 = low) because there is no guarantee of regular transitions between the individual bits. Unless the clocks at the sending and receiving end are synchronised, long strings of 0s or 1s could easily be miscounted. In addition to the need to provide clock recovery, it may be desirable to raise the data transmission rate by using more than two levels for the signal. The conversion from pure binary to a form suitable for transmission at baseband is called *line coding*.

It is useful here to differentiate between the data rate in bits per second and the *signal element* rate. A signal element is the shortest event that can occur in a given coding scheme. The signal element rate, which is also known as the *baud* rate, may be faster or slower than the data rate. If the baud rate is slower than the data rate it means that more than one data bit has been mapped to each signal element, implicitly indicating that more than two levels are being used for transmission. If the baud rate is faster than the data rate then extra bandwidth is being used for transmission and the code must have some other advantage, such as good clock recovery, to compensate for this.

4.4.1 Line Codes

Line codes can be divided into several broad categories:

  (i) non-return-to-zero
 (ii) bi-phase
(iii) delay
 (iv) bi-polar
  (v) multi-level.

They each have advantages and disadvantages, which can be summarised under various headings:

(1) *Bandwidth requirement*: Some codes require more bandwidth for a given data rate than others, depending on how the data bits are mapped to signal elements. Some have a frequency spectrum which extends down to d.c., meaning that they are unsuitable for transformer coupled channels (such as telephone channels).

(2) *Clock recovery*: Some codes are better than others at providing clock-

ing information to the receiver, which is an important consideration for high-speed synchronous data transfer.

(3) *Noise immunity*: Some codes will perform better over noisy channels, that is, they make it easier to discriminate the wanted signal from the noise. Closely related is the ability built into some codes for simple error detection.

(4) *Complexity*: Some codes are more complex to generate and detect, implying more costly transmitters and receivers.

In addition to these variables, there are two properties that must be satisfied before the code is usable:

(a) *Transparency*: the code should be able to cope with any sequence of data bits.

(b) *Uniqueness*: the decoding process must be able to identify uniquely the original binary data.

4.4.1.1 Non-Return-to-Zero Codes

Examples of NRZ codes are shown in Figure 4.4.1. They are called non-return-to-zero because each bit period contains one level only as all transitions are at bit intervals, hence there is no return to a true zero. Information is carried either by the level of the signal during the bit period or by the occurrence of a previous transition. It can be seen that NRZ-L (for level) is the normal binary code: a high level means a logic 1 in that bit period, a low level means a logic 0. NRZ-M (for Mark) uses a transition (low to high, or high to low) at the beginning of a bit period to denote a logic 1, and the lack of a transition to denote a logic zero. NRZ-S (for Space) simply inverts this encoding: a transition means a zero, no transition means a 1.

In terms of the criteria mentioned above, NRZ codes are easy to generate

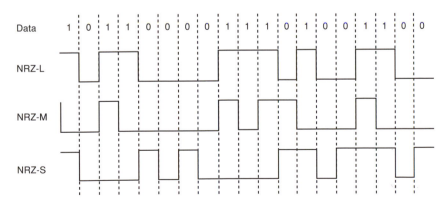

Figure 4.4.1 *Non-return-to-zero line codes*

and detect, but suffer from notable disadvantages. Their ability to transmit clocking information is low, because long strings of 1s or long strings of 0s (depending on the coding scheme) in the data mean that there are no transitions in the signal. NRZ-M and NRZ-S are better than NRZ-L in terms of noise immunity because it is usually easier to detect a transition in a noisy data stream than it is a level. Figure 4.4.2 shows that NRZ codes have a bandwidth which is concentrated between d.c. and about half the data rate, making them very bandwidth efficient. NRZ-L is not uniquely decodable if the sense of the signal is lost, that is, if it becomes inverted then a transmitted 0 will be read as a 1, and vice versa. This might seem an unlikely situation but when twisted pair wiring is used it is very easy to confuse which is the signal and which is the return wire. The system would appear to work correctly except that all the data would be inverted.

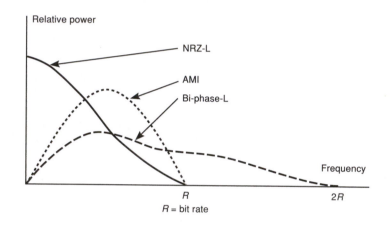

Figure 4.4.2 *Bandwidth requirements for various line codes*

4.4.1.2   Bi-phase Codes

Bi-phase codes all share the characteristic of having at least one transition per bit period. This makes recovery of the clock signal straightforward, indeed they are sometimes known as self-clocking codes. The first shown in Figure 4.4.3 is the Bi-phase-L or Manchester scheme. This uses a high-to-low transition in the middle of the bit period to denote a 1, and a low-to-high-transition to denote a 0. When a 1 follows a 1 or a 0 follows a 0 an extra transition has to be inserted at the start of the bit period, so that the signal is at the required level for the next mid-period transition. Thus the signal element rate could be twice the data rate for long strings of 1s or 0s. Consequently, the bandwidth requirement for bi-phase codes extends out towards twice the data rate, but they have no d.c. level as there is always at least one transition per bit period (see Figure 4.4.2). The major disadvantage of this form of Manchester code is that it is not uniquely decodable. It suffers from the same problem that NRZ-L does, namely that an inadvertent inversion is undetectable and will lead to incorrect data being received.

The other forms of bi-phase coding get around the problem of non-unique decodability, so are more useful. Bi-phase-M and -S are similar in always having a transition at the beginning of the bit period. The -M version uses a transition in the middle of the bit period to denote a 1, with no transition meaning a 0. The -S version is the opposite of this. The Bi-phase-D (for Differential) is also known as the 2nd version of the Manchester code. It always has a transition in the middle of the bit period, with a transition at the start indicating a 0 and no transition indicating a 1.

Systems which use bi-phase codes are more complex than those using NRZ codes, but their other advantages mean that they are in common use. For example, many local area network schemes use a version of Manchester encoding.

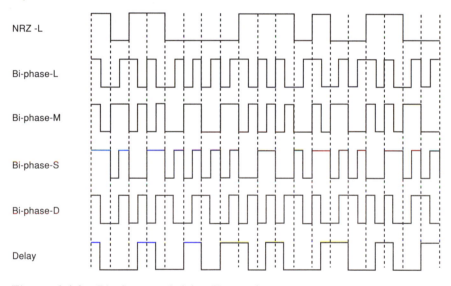

Figure 4.4.3 *Bi-phase and delay line codes*

4.4.1.3 Delay Code

This is also known as the Miller code and is illustrated in Figure 4.4.3. It is related to the Manchester codes but uses fewer transitions. A 1 is encoded by a transition in the middle of a bit period. A single 0 is encoded by no transition, but multiple 0s are split up by transitions at the end of each bit period. Its bandwidth is much narrower than the bi-phase codes but it does have a d.c. component. Under worst-case conditions this residual d.c. component could be significant. The varying rate and position of the transitions make it difficult for a receiver to recover a reliable clock signal. For these reasons Miller coding is not used very often in communications, but it is in common use in data recording, where its narrow bandwidth requirements outweigh the need for a separate clock track.

4.4.1.4 Bipolar Codes

The most common of these is usually known as Alternate Mark Inversion or AMI and is illustrated in Figure 4.4.4. It uses a true zero level to indicate a logic 0, with each successive logic 1 being encoded as alternately positive and negative. This code has no possibility of a d.c. level and many fewer transitions than the bi-phase codes, so it is widely used with channels that have a narrow bandwidth and are transformer-coupled, such as telephone channels. It has no self-clocking facility, but provides some measure of error detection by the need for alternate 1s to be of opposite sign.

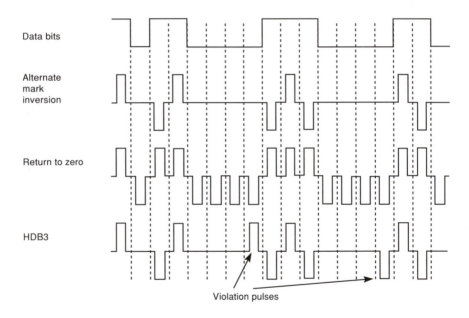

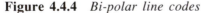

Figure 4.4.4 *Bi-polar line codes*

Variants of AMI are also widely used, perhaps the most common being that known as High Density Binary – no more than three zeros (HDB3). This obeys the rules of AMI except where there are more than three 0s in a row, in which case a pulse, known as a violation, of the same sign as the last mark, is inserted. Subsequent violations alternate in sign from the first one. In order to make the code uniquely decodable, so that violations are not mistaken for genuine marks and vice versa, extra pulses, known as parity pulses, have to be inserted instead of the first zero in the group. HDB3 has the same advantages as AMI but can be considered self-clocking because there are never more than three-bit periods between pulses and bit-synchronisation circuits can cope with short runs without transitions.

Another common bipolar line code is known as 4B3T, where groups of four binary bits are encoded on to groups of three ternary signals (usually known as +1, 0 and −1). As there are 27 possible variants of three signals, each with three levels, it is possible to assign more than one ternary combination to some of the groups of four binary bits. Which combination is used is chosen in order to keep the running sum (adding all the ternary signals together), and hence the d.c. level of the transmitted signal, as close to zero as possible.

4.4.1.5 Multi-level Codes

True multi-level codes, that is, those that have more than one positive and/ or negative value, are also used. One in common use has four levels (usually written as +3, +1, −1, −3). This code is referred to as 2B1Q, with two binary bits being mapped to one of the quaternary levels. It has the advantage of doubling the maximum possible data rate for a given channel bandwidth, but suffers more from noise because the receiver must be capable of distinguishing among four possible levels.

4.4.2 Analogue Transmission of Binary Signals

In many cases the characteristics of the communications channel, for example bandwidth restrictions, make it difficult or impossible to transmit digital data at baseband, however it is encoded. The most commonly quoted example is that of a public switched telephone channel, which is severely band limited to between 300 Hz and 3.4 kHz, restricting the transmission of baseband signals. Other examples are optical fibre channels, which can only pass frequencies in the hundreds of GHz range, or radio wave and microwave channels, which could similarly not support direct digital transmission. In these cases, the binary signal has to be modulated on to an analogue carrier wave. As with the transmission of analogue information, the binary data can be used to modify the amplitude or the instantaneous angle of the carrier. It is common to refer to the various schemes used for transmitting digital data by the terms *amplitude shift keying*, *frequency shift keying*, or *phase shift keying*, because the changes introduced by the modulation process are discrete rather than continuously variable. The common name for a device which performs the modulation and its dual demodulation of binary data is a *modem* (after MOdulation/ DEModulation).

4.4.2.1   Amplitude Shift Keying (ASK)

This modulation method usually involves using one amplitude at a fixed frequency to convey a logic high, and zero amplitude, or the absence of a signal, to convey a logic low. Hence it is often referred to as *on–off keying*. Of course, it is possible to use another amplitude value, rather than zero, to convey a logic low. This is often the way that the signal carried by an optical fibre is modulated, because the laser producing the light can be switched faster if the logic zero takes a residual, or bias, level rather than a true off. It is also possible, although not at all common, to use more than two amplitude values. For example, each of four possible amplitudes could be assigned to one di-bit pair (00, 01, 10, 11), thus each signal element would take one amplitude value but convey two bits of data, approximately doubling the achievable data rate for a given bandwidth. However, the difficulty in receiving this kind of signal in a noisy environment would be significantly greater than conventional ASK.

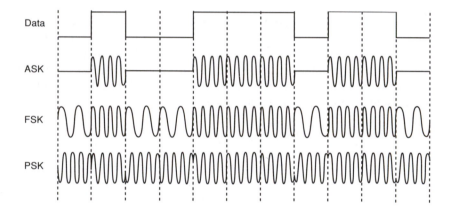

Figure 4.4.5   *Simple modulation schemes*

An ASK analogue waveform $x(t)$, shown in Figure 4.4.5, can be described by the equation:

$$x(t) = a_i \cos [\omega t + \phi]$$

where a is the amplitude function with i taking any integer value (but usually confined to 0 or 1), ω is the carrier frequency and ϕ is an arbitrary phase angle.

A detailed analysis of the bandwidth requirements for ASK is complex, but an easy approximation can be used:

$$B_{ASK} \approx (1 + r)R$$

where B_{ASK} is the bandwidth requirement, R is the signal element rate, and r is the roll-off factor of the transmitter output filters (r will always lie between 0 and 1). Hence an ASK modulated NRZ-L signal of 2400 bps would require a transmission bandwidth somewhere between 2400 and 4800 Hz.

Apart from the case of optical fibres, ASK is rarely used today. It is relatively inefficient in its use of bandwidth compared with phase modulation schemes, and its performance in the presence of noise is not as good. However ASK is easy, and hence cheap, to perform and detect, and is sometimes used if the required data rate over the channel is low (less than 1200 bps).

4.4.2.2 Frequency Shift Keying (FSK)

FSK is also straightforward to describe and perform. One frequency is used to convey a logic high, and another to convey a logic low, as shown in Figure 4.4.5. It can be described by the equation:

$$x(t) = a \cos [\omega_i t + \phi]$$

where i takes the values 0 and 1, and a and ϕ are the fixed amplitude and arbitrary phase angle respectively.

The bandwidth requirement for FSK data can be expressed as:

$$B_{FSK} \approx \Delta f + (1 + r)R$$

where Δf is the difference between the two frequencies used for transmission, and r and R are as defined above. Hence, FSK is less efficient in its use of bandwidth than ASK and is slightly harder to modulate and detect. However, it has been in common use for transmission at low data rates over telephone channels. One international standard divides up the available bandwidth (3 kHz) into two halves, so that full duplex operation can be used. One direction uses frequencies 100 Hz above and below 1170 Hz, while the other uses frequencies 100 Hz above and below 2125 Hz. To avoid the two channels interfering, the bandwidth per channel is restricted to about 600 Hz. Using the formula above, this allows for transmission of data up to about 300 bps.

4.4.2.3 Phase Shift Keying (PSK)

The third property of the analogue wave that can be varied is the phase of the signal with respect to some fixed reference. Pure PSK relies on the transmitter and receiver being perfectly synchronised at all times, so that the reference used by the demodulation process is the same as that used for modulation. This is hard to achieve so it is common to use a scheme called *Differential Phase Shift Keying* or DPSK which uses the phase of the preceding signal element period as the reference. This is illustrated in Figure 4.4.5. As shown, at the points where the data changes from 0 to 1 or from

1 to 0, the phase of the analogue signal reverses. In this case, with only two phases being used, the waveform can be described by the equation:

$$x(t) = a \cos [\omega t + \phi_i]$$

where a and ω are fixed and ϕ_i takes the values 0 or π.

The bandwidth requirements for PSK (or DPSK) are similar to those for ASK, that is:

$$B_{PSK} \approx (1 + r)R$$

DPSK is relatively easy to modulate and detect and is in common use for medium-speed transmission over telephone channels. It is also used for signalling with many satellites, because of its superior noise performance relative to other phase modulation schemes, although it does not offer optimum usage of the available bandwidth.

4.4.2.4 Multiple Phase Shift Keying

If the index i in the equation for phase shift keying is allowed to take more than two values, then multi-level signalling can be achieved. For example, if four phases are allowed, then each of the four di-bit pairs (00, 01, 10, 11) can be assigned to one quadrature phase change, relative to the phase of the carrier wave in the current signal element period. This leads to a doubling of the data rate that can be achieved for a given bandwidth. It is called *Quadrature Phase Shift Keying* or QPSK. However, there is a penalty to pay in terms of noise performance. QPSK is also in common use to achieve higher data rates over telephone channels. This kind of phase modulation can be illustrated using a vector diagram, or constellation, as shown in Figure 4.4.6. The vectors for each di-bit pair are drawn at the appropriate phase angle. Hence, if the next two bits in the data stream are 00, the analogue

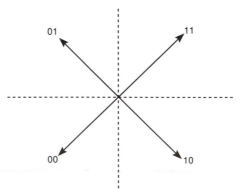

Figure 4.4.6 *Constellation for QPSK*

wave would be made to change in phase by $+45$ degrees. If the next two bits were 11, the phase change would be -135 degrees, and so on.

The distance between the points at which the phases change is called the phase transition interval, and the rate at which phase transitions occur is known as the *baud rate*. This fits in with the previous definition of baud rate, because each phase transition interval represents one signal element.

It is possible to generalise this approach, also known as Multiple Phase Shift Keying (MPSK), by allowing more possible phase changes and hence more bits per baud. If eight phase changes are used (called 8PSK) then three bits per baud can be carried; sixteen phase changes allow four bits per baud, and so on. One common scheme is to use twelve phase changes, four of them having two possible amplitudes. Thus sixteen variations in signal element are permitted and so four bits can be encoded per baud. This is called Quadrature Amplitude Modulation or QAM (see Figure 4.4.7). Now both the magnitude and the direction of the vector imply which quad-bit group is being represented. QAM is used for high-speed data transmission over telephone channels.

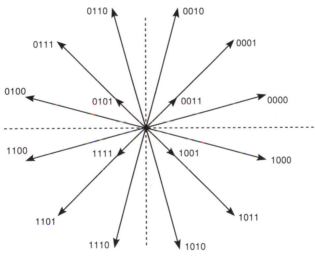

Figure 4.4.7 *Constellation for QAM*

A further step uses a Convolutional Code (see Chapter 3) to encode five or more bits per baud and to give some measure of error control. These systems are commonly referred to as being *Trellis Coded*. The in-built error control allows a higher baud rate for a given channel, because some of the extra errors introduced by lowering the SNR can be corrected. This increase in allowable baud rate for a given residual bit error rate more than compensates for the fact that not all of the bits transmitted over the channel are carrying information. This is known as a *coding gain*.

4.4.3   Detection of Digital Signals in Noise

Whatever modulation method is used, the receiver must be able to distinguish whether a logic 0 or a logic 1 was transmitted in a particular bit interval. The effects of thermal noise will be to make some signals that were transmitted as a 1 appear to the receiver to be a 0, and vice versa. The proportion of times that this occurs in practice is called the bit error rate or BER and can be measured over a statistically significant period of time.

Figure 4.4.8 illustrates how thermal noise could cause the receiver to make an error. The simplest case is shown, that of straight binary transmission or NRZ-L. The incoming signal will have a value around that of the transmitted bit with a statistical variation described by the probability density function $p(V)$ of the noise. The receiver will apply a threshold to the incoming signal, taking anything over the threshold to be a 1 and anything below it to be a zero. The shaded area in Figure 4.4.8 shows that some bits will be determined in error if the received value of a signal element transmitted as a 1 or 0 can be taken over the threshold by the addition of noise. How much the curves illustrating the possible variations in received signal overlap is determined by the signal-to-noise ratio of the channel.

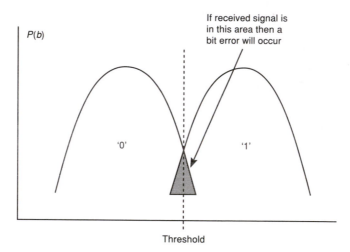

Figure 4.4.8   *Probability of a bit error occurring*

It is shown in Appendix 1 that for Gaussian distributed noise the error probability can be calculated using the Q function defined by:

$$Q(x) = \frac{1}{\sqrt{2\pi}} \int_x^\infty \exp\left(\frac{-t^2}{2}\right) dt$$

$Q(x)$ is a tabulated function, representing cumulative values of the Gaussian distribution. For a simple unipolar binary signal with levels 0 and x, and a decision level $x/2$, the probability of a bit error is given by $Q(x/2)$. Some representative values are given in Table 4.4.1.

Table 4.4.1 *The Q function*

x	$Q(x)$	x	$Q(x)$
0	0.5	1.28	0.1
0.5	0.309	2.33	0.01
1.0	0.159	3.10	0.001
1.5	0.067	3.72	10^{-4}
2.0	0.023	4.27	10^{-5}
3.0	1.35E-3	4.76	10^{-6}
4.0	3.17E-5	5.20	10^{-7}
5.0	2.87E-7	5.60	10^{-8}

The signal-to-noise ratio of the channel at the receiver is usually the factor that determines the ease with which the signal can be recovered. However, for a digital signal it is the energy per bit and how much bandwidth is used, rather than the average signal and noise powers, that are the crucial factors in determining how much of a problem the noise might cause. Hence, the ratio of signal-power-to-noise-power is better described by the ratio of signal energy per bit to thermal noise power per Hz (E_b/N_0) where:

E_b ($= S/R$) is the ratio of the signal power S to the data rate R (not the signal element rate)

N_0 ($= N/B$) is the noise power per Hz, where N is the total noise power and B is the bandwidth of the channel.

$$\frac{E_b}{N_0} = \frac{\text{signal power } S \text{ / data rate } R}{\text{noise power } N \text{ / bandwidth } B} = \frac{S}{N} \times \frac{B}{R}$$

Hence E_b/N_0 is a measure of the worth of a channel used for digital signals, just as SNR is a measure of the worth of a channel used for analogue signals. Note that no account is taken of the effects of cross-talk or impulse noise, as these are either avoidable or unpredictable.

It can be shown that, for unipolar signaling, BER and E_b/N_0 are related by:

$$\text{BER} = Q\left(\sqrt{a\frac{E_b}{N_0}}\right)$$

where a is a constant determined by the line coding scheme. For the simple unipolar binary signal analysed in Appendix 1 it is shown that the error rate is given by $Q(S/2N)^{1/2}$. Putting $a = 1$ in the above formula and signalling at the Nyquist rate $R = 2B$ gives the same result.

To give an example of how BER is determined in a practical application, let us assume a channel with S/N of 10 dB, a bandwidth of 3 kHz and a data rate of 2400 bits/s. The coding scheme in use is NRZ-L which makes the parameter a have a value of 1. Hence:

$$\frac{E_b}{N_0} = \frac{S}{N} \times \frac{B}{R} = 10 \times \frac{3000}{2400} = 12.5$$

The BER is therefore given by Q (3.54) which is about 2×10^{-4}, or in words, on average two bits out of every ten thousand will be received in error. This would be unacceptable in most applications, so the BER would have to be lowered by raising the channel signal-to-noise ratio (by boosting the source power), or by lowering the data rate. In comparison with the ideal theorem, $C = B \log(1 + \text{SNR}) \approx 10\,000$ bits/s with no net errors.

These calculations only take thermal noise into account, because the effects of cross-talk and impulse noise are much more difficult to quantify. Hence they should be considered to be producing estimates of the BER, which will be lower than the true figure. For most applications, the difference between the estimate produced by these calculations and the true BER will be negligible. However, in some channels, such as satellite links, the effects of impulse noise could make the estimated figure for BER unrealistically optimistic.

Other line coding schemes make the constant a take a value between 0.5 and 2.0, depending on the ratio of the baud rate to the data rate. If binary modulation of analogue signals is used, then for FSK a is 1, and for PSK a is 2, implying that PSK is much better than FSK in terms of noise performance. For multi-level signalling, such as QAM, it is much more complex to evaluate the probable BER. This is because the discrete elements that may be received in error are no longer individual bits, but groups of bits. For example using QPSK, if a received signal element is assigned to the wrong di-bit because of the effects of noise, how do you know whether one or both bits are in error? The original assignment of di-bits to phase changes can play a part in minimising the possible errors, for example by making sure nearest neighbours only differ in one bit position. This would make the bit error rate equal to the signal element error rate. However, this way of assigning codes may be more difficult to achieve when there are more than two bits per signal element. Then the relationship between signal element error rate and bit error rate becomes a further probabilistic equation, including a dependency on the relative likelihood of transmission of each signal element.

If the codes are assigned such that nearest neighbours only differ in one bit position then it can be said that:

$$\text{BER} = \frac{P_E}{k}$$

where P_E is the probability of a signal element error and k is the number of bits per signal element. Figure 4.4.9 shows the variations in predicted BER against E_b/N_0 for differing values of k.

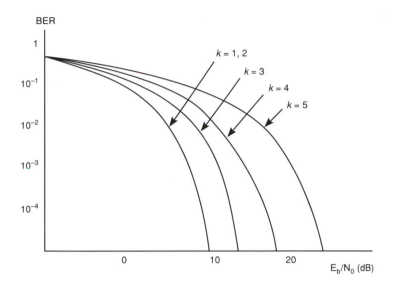

Figure 4.4.9 *Bit error rates for multiple phase shift keying*

It can be seen that as k increases the predicted BER rises. This is as expected because of the smaller differences between signal elements as k gets larger. This might seem to argue against the use of multiple phase shift modulation techniques, but all that is happening is the classic communications trade-off between efficient use of bandwidth and noise performance.

4.5 Examples for Solution

1. (i) How does the power in an AM wave depend on the modulation index?
   (ii) Is any information lost if only one AM sideband is transmitted? Why are both sidebands usually transmitted?
   (iii) How does the power in an FM wave depend on the modulation index?
   (iv) How do the spectra for AM and FM differ if β is very small?
   (v) What are the main advantages and disadvantages of FM compared with AM?
2. Discuss the feasibility of the following systems from the point of view of Information Theory, comparing them with conventional analogue systems:

   (i) A solid-state substitute for a tape recorder, employing no moving parts but comprising only logic and storage elements, for low fidelity digital recording of speech. Assume that the maximum storage is 10^6 bits.

   (ii) A solid-state still-frame colour camera (that is, equivalent to a typical 35 mm camera) in which the light is detected by a c.c.d. chip and stored in RAM. The pictures taken are to be viewed later on a TV screen and converted to more permanent storage as required. Assume that the total storage available is 10 Mbit, and state any other assumptions made.

3. An electronic 'tune generator' can produce eight independent notes, the lowest having a frequency of 256 Hz and the highest 512 Hz. The notes may be of duration 0.2, 0.4, 0.8 or 1.6 s; a rest (silence) of the same durations may also be produced.

   (i) Find the average information produced per second, assuming all rates and durations equally likely, and hence estimate the number of different 'tunes' of duration 5 s.

   (ii) The 'tunes' are to be transmitted via a 3 kHz telephone channel. For each of the following systems find the bandwidth needed and comment on the feasibility:

     (a) double-sideband AM with $m = 1$

     (b) FM with $\beta = 5$

     (c) PCM with 256 levels

     (d) a tone-coding method in which a binary code is assigned to each tone or rest specifying its frequency and duration.

4. The binary sequence 100110000010111 is to be transmitted over a communications channel. Sketch the waveforms for the following scenarios:

   (i) transmission at baseband using differential-Manchester line coding

   (ii) transmission at baseband using HDB3 line coding

   (iii) transmission over a telephone channel using QAM modulation.

5   Applications

5.1   Introduction

The developments in technology over the last few years have greatly increased the practical applications of Information Theory. However, technological progress has been so rapid that many would-be communications or information engineers have been unable to develop the necessary theoretical background on which real understanding and progress depend. An understanding of the principles of Information Theory is essential in order to exploit the great potential in practical applications that are now available.

We will not attempt a survey of the applications of Information Theory. Such 'overviews' tend to become simply a list of words; they are easy reading and may enable one to follow some of the current technological jargon but add little to one's real knowledge of the subject. Instead we will consider the properties and applications of some important practical information channels, such as video, telephone and audio channels, and then discuss some of the interesting recent developments such as CD technology and local area networks.

5.1.1   Typical Information Channels

The main means of telecommunication nowadays are, of course, via radio or cable. Radio communication systems include television and sound broadcasts, cellular telephones, navigation systems, time broadcasts, microwave links, wireless telegraphy etc., and many special-purpose systems (spacecraft communications, emergency services etc.). Cable systems involve optical fibres/coaxial cables and 'twisted pairs'. The former provide channels of wide bandwidth suitable for television, radio or many multiplexed telephone channels, whereas the latter are low bandwidth channels for data or telephones. Optical fibres have now replaced coaxial cables for many applications on account of their greater bandwidth, smaller diameter and freedom from electrical interference.

Television and telephone channels are particularly interesting, partly because they can be analysed quantitatively but also because of the important developments in information services, including teletext (Ceefax and Teletext) and facsimile systems. The feasibility of these systems originally depended very much on the fact that almost everyone has ready access to television and telephone receivers, but the more recent developments were very much dependent on the availability of cheap yet powerful digital processing equipment. A rapidly increasing number of people now have access to the

Internet via a PC, and this in turn is already having a very large effect on information services.

5.2 Television Channels

In the UK, a television picture is built up from 625 horizontal 'lines', though not all the lines are actually displayed, one complete picture (one frame) being transmitted 25 times per second. A technique known as 'interlaced scanning' is used, whereby the odd-numbered lines (one 'field') are transmitted in 1/50 s, followed by the even-numbered lines in the next 1/50 s; this avoids the flickering that would be apparent if all the lines were sent together 25 times/s, and uses only half the bandwidth required by a flicker-free 50 frame/s system.

The form of the signal for one line is shown in Figure 5.2.1(a). A precisely timed synchronisation pulse precedes the 'colour burst', containing the colour information, and the brightness information follows in the 52 μs interval shown. The spectral range of the signal is shown in Figure 5.2.1(b). Most of the bandwidth is taken up by the luminance information, the chrominance signal requiring much less bandwidth (since the colour in a typical picture changes more slowly than the brightness). The basic signal has a bandwidth of about 6 MHz, and is transmitted using vestigial-sideband-modulation (as described in Chapter 4), so the channel bandwidth required is only 7.5 MHz instead of 12 MHz for standard AM. The spacing between channels is defined as 8 MHz. The audio bandwidth is negligible in comparison, and is transmitted by FM.

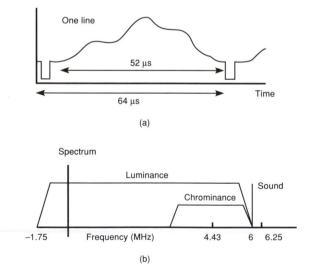

Figure 5.2.1 *Television signal (a) and its spectrum (b)*

5.2.1 Information Content

Since the colour information is relatively low we will consider a monochrome picture with height h and width w, as in Figure 5.2.2. The vertical resolution is $h/625$ and it is reasonable to assume a similar horizontal resolution, dividing the picture into many small square elements (picture elements or 'pixels'). The bandwidth used is actually chosen to make this approximation valid.

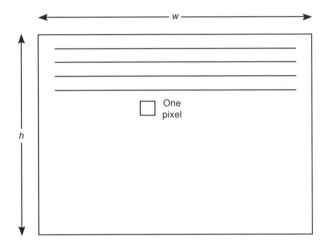

Figure 5.2.2 *Pixels approximation to television picture*

A typical ratio of w to h is 4/3 so the total number of pixels is $625 \times 625 \times 4/3$ and the number transmitted per second about 1.3×10^7. Using the usual $2B$ rule this requires a bandwidth of 6.6 MHz. In practice, some time has to be allowed between lines and frames for synchronisation, as shown in Figure 5.2.1(a), only 575 being used for the actual picture (some of the remainder are used for teletext).

The total information in one frame is equal to the number of pixels times the information per pixel. The latter depends on the number of distinguishable shades, which in turn depends on the signal-power-to-noise ratio. If we assume that 32 shades can be distinguished, the information per frame becomes $625 \times 625 \times 4/3 \times \log(32) \approx 2.6 \times 10^6$ bits equivalent to 6.5×10^7 bits/s (usually denoted 65 Mbps). This is quite a lot of bits and it is only recently that the resolution available in a typical PC screen has approached this value.

The theoretical capacity depends on the power-to-noise ratio. Assuming a value of 1000 we obtain $C = 8 \times 10^6 \log(1000) \approx 80$ Mbps, in close agreement with the above value. However this excellent result is because we have cheated in making the various assumptions, though it was done somewhat subtly. We saw above that the Ideal Communication Theorem can be derived approximately by assuming that the number of distinguishable levels is equal to the root mean square signal-to-noise ratio, and our chosen number of levels (32) is very near to the square root of our assumed power-to-noise ratio. However, the assumptions are very reasonable and we can therefore fairly accurately quantify the information in a television picture.

The wide bandwidth needed for transmitting analogue TV signals means that the number of channels available to broadcasters is extremely limited, given a fixed allocation of the overall frequency spectrum for broadcast TV. There has been a lot of research and development effort into techniques for reducing the bandwidth per channel, mostly centred around using digitised versions of the standard TV picture. The most common approach is to take 16-bit samples of the TV signal at a rate of 13.5×10^6 per second (that is, at about the $2B$ rate), producing 216 Mbps. As the video signal is now in digital form, lossy compression techniques can be used to reduce the data rate needed to convey the essential information. The MPEG-2 standard (outlined in Chapter 3) has been adopted as the most appropriate compression mechanism, and it can reduce the data rate needed to transmit a normal TV picture (and its associated sound) to around 6 Mbps. Qualitative tests are used to determine the acceptability of the picture so it is impossible to express a precise relationship between the information content and the final data rate.

In Europe, it is proposed that a technique known as *Coded-Orthogonal Frequency-Division-Multiplexing* (COFDM) will be used to carry the digital signals for transmission. COFDM embeds a pseudo-random code into the data and then spreads it out over a large number of carrier frequencies. Crucially, this technique allows several information signals to share the same transmission bandwidth. The receiver uses sophisticated digital signal processing to recover individual signals by locking on to the appropriate pseudo-random code. The US are proposing to use multi-level phase-amplitude techniques (see Chapter 4) instead of COFDM. In either case, there is the possibility of broadcasting up to six digital signals in the bandwidth occupied by one conventional analogue channel. The spreading-out of the signal and use of digital processing for recovery can also allow digital signals to be overlaid in the same frequency band as analogue signals. The analogue television pictures appear as noise to the digital decoders and are rejected. Hence, the transition from analogue to digital TV can be achieved without the need for new bandwidth being allocated for broadcast use. Up to 18 terrestrial and 200 satellite channels are being planned. How the programme makers will find enough material for all these channels is, thankfully, not our concern. Digital TV broadcasting, both terrestrial and from a satellite, is scheduled to start in 1997 or 1998.

Alongside the development of digital techniques to reduce the bandwidth needed to transmit standard-quality TV pictures, has been a drive towards higher quality, known as *High-Definition TV* (HDTV). This would increase the vertical resolution of each picture to 1250 lines, and change the aspect ratio of the picture from 4:3 to 16:9. Standards for HDTV have been slow to arrive because of a competition between those who favour the analogue route to ease transition and those who wish to go straight to digital. At the time of writing it seems that the proponents of the digital method have won. This scheme (using sampling and MPEG-2) will result in a compressed bit-stream at around 24 Mbps, so one HDTV channel will have to be traded for four conventional-quality channels. Technology is now losing out to marketing and no dates have been set for the widespread broadcast of digital HDTV pictures, although some test broadcasts were made, using all analogue technology, of the 1992 Olympics from Barcelona.

5.2.2 Teletext

The UK has always been at the forefront in Information Technology, and was the first country to provide public information services in Teletext and Viewdata. Teletext transmissions began in 1976 and Viewdata (British Telecom's Prestel) started in 1979 though it has now been withdrawn. Confusingly, teletext refers to both the technology and to a private company providing the service to ITV. The BBC teletext service is given the name Ceefax. Both systems provided a wide variety of information to the user for display on a television screen; in teletext the information is transmitted with the normal TV picture, but in Viewdata the TV set was linked to a telephone. The systems were developed together using the same screen format, one page comprising 24 rows of 40 characters. Some graphics characters are included, employing a 2 × 3 matrix as shown in Figure 5.2.3, so that the screen resolution for graphics is 80 × 72. This is fairly coarse, though seven colours are available, but has the particular advantage that the screen data can be stored in only 1 kbyte of RAM, since an 8-bit ASCII code is used.

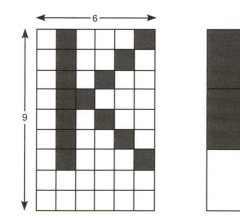

Figure 5.2.3 *Teletext screen*

The teletext system uses four otherwise-unused lines per frame, two in each interlaced scan. The data are digitally encoded and each line carries the data for one 40-character row so one page takes 6/25 s and about four pages are transmitted per second. The data rate is easily calculated, since 45 8-bit bytes (40 bytes for characters, plus 5 synchronisation bytes) are transmitted in the active line time of 52 μs, giving a rate of 6.9375 Mbps. A teletext data signal is shown in Figure 5.2.4.

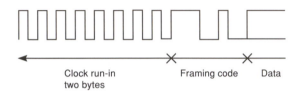

Clock run-in   Framing code   Data
two bytes

Figure 5.2.4 *Teletext data signal*

The system library comprises 800 pages, divided into 100-page 'magazines' so the cycle time is nominally about 25 seconds. However, some of the more popular pages are included more than once per cycle so the access time is somewhat reduced. In addition, each page may have up to 3200 sub-pages (making about 2.5 million in all) but the access times become very long since only one sub-page step would be available per 100-page cycle. The travel information is often in this form, and by the time one has waited for a complete cycle one could often have nearly reached one's destination!

The alphanumerics, graphics and control characters are all represented by a 7-bit ASCII code plus one parity bit. Single errors are detected and the character ignored, so that some gaps are left in the received page on its first reception. However, double errors can be very serious, especially if they occur in a control character, so that a whole line of text could be changed in colour or into graphics characters. This possibility is reduced in headings, magazine and row addresses, by employing an (8, 4) Hamming code to provide single-error correction and double-error detection, characters with detected double errors being ignored. The format of the page header and the subsequent lines is shown in Figure 5.2.5.

The Prestel viewdata system had the advantage that it was a two-way interactive system so that the library was not essentially limited to the 800 pages of teletext, but only by the size of the controlling host computer. Terminals were located in public places, such as libraries, but it never gained widespread usage. The increasing availability of PCs linked to the Internet via modems further reduced the value of the system and it has now been withdrawn. It is interesting to note that the French Minitel system, while technologically very similar to Prestel, has been highly successful. Its original purpose of replacing paper-based telephone directories led to the widespread

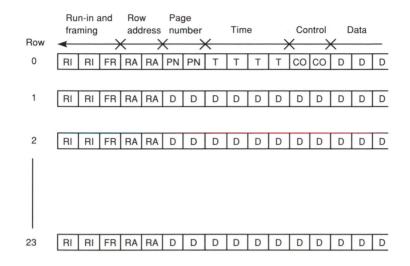

Figure 5.2.5 *Format of Teletext lines*

availability of terminals (every home was given one), so a virtuous circle was started of more use leading to more information being made available, leading to more use and so on. It is rumoured that the most profitable sections are the so-called pink pages, where sexual services for all tastes (including many that the authors would not dare to translate) are available.

5.3 Telephone Channels

Surprisingly, telephone channels were actually developed for people to speak through, though this is much the least efficient and most wasteful way of using them nowadays. In the early development of the telephone it was found that speech requires a bandwidth of about 3 kHz for reasonable quality and to retain the characteristics of the speaker's voice; with lower bandwidths the intelligibility rapidly falls and everyone tends to sound the same (perhaps some sociologists or educationalists would recommend a reduction). The actual bandwidth is taken as 300–3400 Hz, though for practical purposes this is usually rounded to 3 kHz and sampling is typically at 8 kHz for digital applications to simplify recovery.

5.3.1 Information Content

Assuming a power-to-noise ratio of 100 (r.m.s. signal-to-noise ratio 10) the ideal capacity is 3000 log(100) ≈ 20 000 bits/s (20 kbps). Speaking at a rate of two five-letter words per second provides about 10 bits/s (50 bits/s, ignoring redundancy) which does not compare too well. Speaking is clearly the world's worst way of conveying information; it can be calculated that if the authors had been able to use the bandwidth of their voices efficiently, the information in their lecture course on Information Theory could have been given in only 30 seconds!

The problem lies in the way in which the speech information is carried and the relatively large bandwidth required to communicate it. This was realised many years ago, and many attempts have been made to improve the situation. It appears that the information in speech is carried more by the low-frequency modulation of various 'carrier frequencies' in the waveform than by the 'carriers' themselves. When a particular sound is spoken a particular set of harmonics is produced, but it is the modulation of the individual harmonics that is most important. Figure 5.3.1 gives a typical speech waveform and the corresponding 'spectrogram', showing how the frequency components vary with time. Since one usually has to transmit most of the harmonics a large bandwidth is required, whereas if one could extract the modulation itself a considerable saving would result.

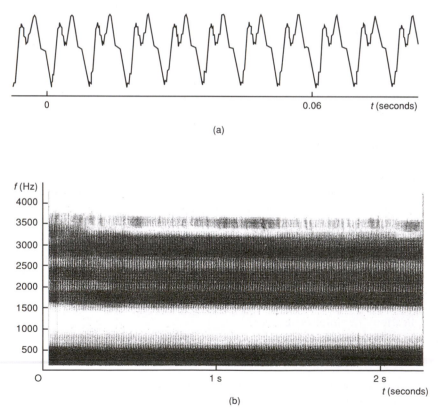

Figure 5.3.1   *Speech waveform for the sound 'EEEE' (a) and its spectrogram (b)*

This is a simplification, of course, but a system based on this idea was demonstrated as long ago as 1936 by Dudley in his vocoder (Dudley, 1936) and is shown in Figure 5.3.2 on page 172. The speech was divided into a number of bands, mostly 300 Hz wide, and the low-frequency modulations in each band extracted by envelope demodulation using diodes followed by 25 Hz low-pass filters, as in AM. One of the channels was used to detect whether the sound was predominantly a 'buzz' or a 'hiss' and transmitted its frequency in the former case. The overall bandwidth was thus reduced to 250 Hz from the original 3 kHz. The ten channels could be multiplexed together and transmitted to the receiver, using SSAM or other methods, and using a bandwidth still of the order of only about 300 Hz. The receiver comprised a similar set of bandpass filters, the inputs being controlled by the 'buzz/hiss' channel. A noise generator was switched in for a 'hiss' and a square wave of appropriate frequency for a 'buzz', the 25 Hz modulation signals being used to modulate the appropriate frequency band. Dudley's vocoder produced intelligible speech, though all speakers sounded much the same. It was a remarkable achievement at the time, and even with the technological developments now available it is difficult to obtain a bandwidth reduction of better than a factor of 10.

The distribution and characteristics of the harmonics present in speech clearly have a critical effect on the sound produced. One of the authors has an interesting theory on the subject. It is a truth not yet universally acknowledged; in fact he is unaware of anyone else who admits to believing it (including the co-author), though for the last 25 years he has been the world's leading proponent. The Harmonic Emotional Theory* states that specific harmonics or groups of harmonics in a speaker's voice produce specific emotional effects. The theory developed from the author's critical observations of his children. It is well known that small children are binary creatures – they are either sleeping or crying. In the former case they are only mildly unpleasant but in the latter they are particularly nasty and require either feeding or changing. The author noted that his older child's crying was quite soothing and could easily be slept through. However, at the first cry of the younger child he was instantly awakened and had to be physically restrained from attacking it (he did not attempt to feed or change it). Clearly a particular set of harmonics were inducing an uncontrolled emotional response. Even more convincing is the author's observations on students attending his lectures. It is clear that the harmonic content of his voice has a pronounced soporific effect on students! They fall asleep within a few minutes but awaken as soon as the lecture is over. There appears to be no other explanation than the Harmonic Emotional one**. The recent outbreaks of 'road rage' have a similar explanation – it's all to do with the tone of one's horn.

* T-shirts with the logo 'I am a HE man' will be available if demand is sufficient.
** This has led to an interesting definition of a lecturer as 'someone who talks in other people's sleep'.

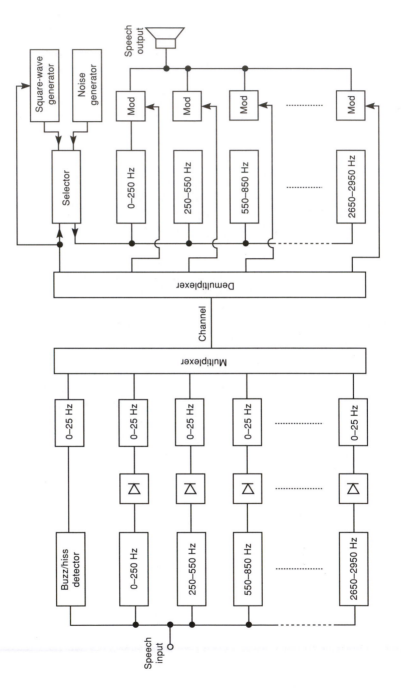

Figure 5.3.2 *Dudley's vocoder*

Most telephone companies now carry all information in digital form, except over the *local loop* from user to local exchange. Typically speech is sampled at 8 kHz with 8 bits per sample, leading to a base data rate of 64 kbps per conventional telephone channel. Much research work has gone into techniques for reducing this rate while preserving the information content. One that is widely used is known as *Codebook-Excited Linear-Predictive* (CELP) coding. This was developed at AT&T Bell Labs in the mid 1980s. In its original form it took a supercomputer about 100 seconds to encode 1 second of digitised speech. However, advances in signal processing integrated circuits have made it possible to implement the technique in real time with quite small amounts of hardware. The basis of the algorithm is to divide the waveform up into short (say a few tens of milliseconds) bursts and to use predictive coding on each burst. This is a technique where the most likely value of the next sample is predicted, using a general model of speech. This generates a set of coefficients that can be transmitted instead of the waveform. The real waveform is also compared with that which the predictor will produce at the output and the set of differences between the two is generated. If the prediction is a good one then the differences will be much smaller than the waveform values themselves. These differences can be compared with sets of stored differences and the address of the closest set sent. In addition, it is known that the coefficients for any prediction are correlated and can be generated from an analysis of the first few predictions. Then only a look-up table address need be transmitted, rather than all the coefficient values themselves.

So for each burst the encoding circuits generate some predictor coefficients, a look-up table address for the rest of the coefficients and a codebook address for the differences. This allows a reduction of about 10 to 1 in the number of bits to be transmitted, meaning for example that a 64 kbps telephone channel could carry ten digitised voice channels rather than the one that would be possible if no encoding was used.

Telephone channels are now commonly used with a modem to send data between computers. The modulation techniques most commonly used have all been described in Chapter 4. Modern modems, using Trellis coding, can achieve data rates of 28.8 kbps over the 3 kHz bandwidth without any compression being employed. As modems are usually implemented using digital signal processing, it is relatively easy to add sophisticated functions such as error correction and lossless compression. It is important that methods for transmission of data are standardised so that you do not have to guarantee that the receiver is using the same brand of modem as the transmitter. The *International Telecommunications Union* have a standardisation committee (known as ITU-T) which specifies how modems should operate. These standards are all given numbers preceded by a letter *V*. So a *V*.34 modem will support 28.8 kbps without compression and will offer *V*.42 error correction and *V*.42$_{bis}$ data compression. Error correction is performed by dividing the data up into packets or frames, adding an error-detecting code (a CRC code – see Chapter 3) to each frame and re-transmitting any frame received with detected errors. Data compression uses an LZ type algorithm (see Chapter 3) to give compression rates up to 4:1. Any manufacturer's *V*.34 modem should interoperate with any other and, in most cases, they do.

Using analogue transmission for digital data over the home-to-exchange link, followed by sampling to convert to a new form of digital data at the exchange, is wasteful of resources and leads to higher than necessary bit-error-rates, because of the number of conversion steps needed. The telephone companies recognised this and have specified a digital access mechanism to their systems, known as the *Integrated Services Digital Network* (ISDN). A full description of ISDN is beyond the scope of this book but the ideas behind it are simple. The user gets access to a fully digital (64 kbps) channel, known as a *B-channel*, which only carries data. The signalling associated with this data is carried by a separate channel (operating at 16 kbps), known as a *D-channel*. Telephone companies have agreed that these channels will be provided to users in groups, rather than individually. So a *Basic-Rate* ISDN service will provide two B-channels and one D-channel. All are time-multiplexed down the same physical connection, which is commonly the same copper wiring used to provide an analogue telephone service. For businesses a *Primary-Rate* service is offered, with 30 B-channels and one (64 kbps) D-channel. Fibre-optic or good-quality coaxial copper cable must be used to carry these logical channels, which, including framing bits, give a total rate of 2.048 Mbps. Much higher rates are promised by the *Broadband* ISDN (B-ISDN), with users getting access to rates up to 600 Mbps. A date when the B-ISDN (which is the true *Information Superhighway*) will be offered by the telephone companies has not yet been specified.

5.3.2 Facsimile Transmission

An interesting application of a telephone data channel is facsimile (FAX) transmission of pictorial data. Analogue facsimile systems were developed during the Second World War. A document or photograph was scanned optically, in much the same way as a television picture is produced. A scanning rate of 3 lines/s with a vertical resolution of 4 lines/mm produced about 2500 pixels/s for a 210 mm wide page, assuming the page to be divided into 'square' pixels of side 1/4 mm. This required a bandwidth of 1250 Hz so double-sideband AM could be used for transmission via a standard telephone channel. The time taken for a page of A4 size (approximately 210 mm × 300 mm) was about 5 minutes, so the system was somewhat limited in its application, though telephone calls were relatively cheap in those days

The original analogue system was known as 'Group 1 FAX'. The most widely used system at present is a digital system known as 'Group 3 FAX'. A page is scanned in monochrome at a resolution producing 1188 pixels per

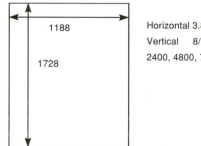

Horizontal 3.85/mm
Vertical   8/mm
2400, 4800, 7200, 9600 baud

Figure 5.3.3 *Pixels in FAX document*

horizontal line with 1728 lines as in Figure 5.3.3. For a page of A4 size this corresponds approximately to a resolution of 3.85 pixels per mm horizontally and 8 pixels per mm vertically. If the pixels were to be transmitted directly then one page would produce $1188 \times 1728 = 2.05 \times 10^6$ bits, which would take about five minutes via a standard telephone channel. The data compression method used employs a system known as run length encoding. Each line is broken down into runs of black and white pixels, and each run expressed as two codewords. The first codeword (known as the 'make-up' codeword) is the number of complete multiples of 64 in the run, and the second (the 'terminating' codeword) is the additional length from 0 to 63. An end of line (EOL) code is used to signify that the rest of the line is all black or all white. The different run lengths are then coded by a method known as 'modified Huffman', which is a compact code similar to the standard Huffman method in which the more probable runs are given shorter binary codes. Some of the standard codes are shown in Table 5.3.1.

Table 5.3.1 *Modified Huffman codes used in FAX*

Make-up codewords			Terminating codewords		
Run	White	Black	Run	White	Black
64	11011	0000001111	0	00110101	000110111
128	10010	000011001000	1	000111	010
192	010111	000011001001	2	0111	11
256	0110111	000001001011	3	1000	10
320	00110110	000000110011	4	1011	011
384	00110111	000000110100	5	1100	0011
448	01100100	000000110101	6	1110	0010
512	01100101	0000001101100	7	1111	00011
576	01101000	0000001101101	8	10011	000101
. . .			. . .		
. . .			. . .		
1728	010011011	0000001100101	63	00110100	000001100111
EOL	000000000001				

The codes used were deduced by analysing typical documents, and produce an overall compression by about a factor of 20. Transmission is typically at baud rates of 9600, 7200, 4800 or 2400, and at the lowest rate the transmission time is $(1/20) \times 2.05 \times 10^6/2400$ or about 40 seconds.

5.4   Audio Channels

Audio channels are mostly used for transmitting music. Radio broadcasts use FM to obtain a higher signal-to-noise ratio at the receiver than would be possible with AM and have a signal bandwidth of typically 15 kHz (the bandwidth on the channel is therefore about 180 kHz with $\beta = 5$). Assuming a signal-to-noise ratio of about 1000 the theoretical channel capacity can be calculated as $15\,000 \times \log(10^6) \approx 300$ kbps. Unfortunately there is no easy way (or even a difficult way) of estimating the actual information content of music in, say, a radio broadcast. It probably varies considerably with the type of music in a similar manner to English text, so that the highly unpredictable modern music (so-called) would have a higher information content than music providing some degree of enjoyment. However, when the signal waveform is digitised, as in the recording of music on CDs or other digital media, one can at least estimate the data content, if not the actual information content of the music.

5.4.1   Compact Discs

Standard compact discs (Figure 5.4.1) are 12 cm in diameter and store the information in a spiral track of pits, varying in length between 0.9 μm and 3.3 μm, with pitch 1.6 μm. They employ constant linear velocity scanning at about 1 m/s, and the rotation rate varies between 200 and 500 revolutions per minute.

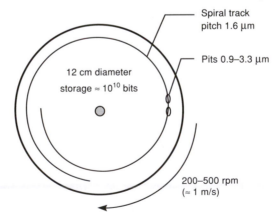

Figure 5.4.1   *Compact disc*

The audio signal bandwidth is 20 kHz and the sampling rate is standardised at 44.1 k samples per second (there is apparently a logical reason for this number). Stereo samples are taken with a resolution of 16 bits, providing a data rate of 176.4 kbytes/s (1.411×10^6 bits/s). The data are divided into frames (one frame is 24 bytes, comprising six left and right 16-bit samples), and blocks (comprising 98 frames); there are a total of 270×10^3 blocks, so the total storage is $24 \times 98 \times 270 \times 10^3 \approx 600$ Mbytes (about 10^{10} bits). The playing time is 74 minutes. From the discussion of quantisation noise the signal-power-to-noise ratio is equal to the square of the number of levels (2^{16}) so the r.m.s. signal-to-noise ratio is $20 \log_{10} (2^{16}) = 96$ dB. This is usually referred to as the dynamic range, which is strictly the ratio between the largest r.m.s. signal and the r.m.s. noise.

Encoding and Decoding

Compact discs use a coding system known as Cross Interleaved Reed–Solomon (CIRS), which is a cyclic block code with interleaving to reduce burst errors. The code operates on one frame of 24 bytes, adding four error-correcting bytes in a first stage and a further four bytes in a second stage, producing overall a (32, 24) byte block code.

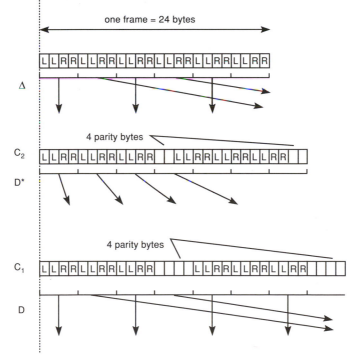

Figure 5.4.2 *Encoding process for CDs*

The encoding process is very complex and is shown in Figure 5.4.2, and involves the following steps:

(a) Δ interleave. Even-numbered samples are separated from odd-numbered samples by two frame times.
(b) C_2 encode. Four RS parity bytes are added, comprising the 'inner' code.
(c) D^* interleave. Each byte is delayed by a different length.
(d) C_1 encode. Four more RS bytes are added, comprising the 'outer' code.
(e) D interleave. The even bytes of one frame are interleaved with the odd bytes of the next.

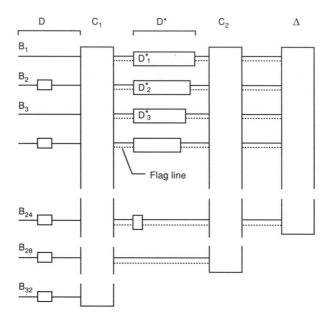

Figure 5.4.3 *Decoding process for CDs*

The decoding process involves the detection and correction of errors and the reassembly of the data, again requiring five steps and shown in Figure 5.4.3.

(a) D de-interleave. The 32 bytes are applied in parallel with alternate 1-byte delays.
(b) C_1 decode. This stage can correct one byte and can detect longer bursts. Multiple errors are flagged on all 28 remaining bytes, after stripping the four parity bytes.
(c) D^* de-interleave. The different length delays in the D^* process cause errors in one frame to be spread over several frames, so the C_2 decoder can correct them.

(d) C_2 decode. This stage attempts to correct the burst errors flagged by C_1. If there are too many for correction the bytes are flagged and passed on.
(e) Δ de-interleave. De-interleaves uncorrectable but detected byte errors, allowing interpolation to be used between reliable neighbouring samples.

The error detection and correction is very effective overall. The maximum length of burst correction is 16 consecutive frames, after which interpolation between samples is used, followed by muting for bursts of more than 48 frames. These values correspond to about 2 mm and 5 mm respectively on the disc, though the authors have not yet persuaded their students to try the appropriate experiments on their own CDs. New discs have a binary error rate of about 10^{-4} and the probability of an undetected random error is negligible.

5.5 Computer Networks

It is now very common to connect computers together in a network, to enable information (data or programs) or resources (for example, access to printers) to be shared. If the computers are physically close together, a *Local Area Network* (LAN) can be used to provide the physical channel over which the information will travel. Increasingly, the use of fibre-optic cables means that the word *Local* now refers to a style of interconnection rather than a distance. LANs extending up to 100 km in size are now common. If the computers are not local (possibly scattered world-wide) then a *Wide Area Network* (WAN) can be used. Unfortunately, there are no clear distinctions that help to classify a particular network as either a LAN or a WAN. The easiest distinction can be made by deciding who will provide the physical interconnections. If it is you, the user, then you are installing a LAN. If you buy or lease the carrying mechanism from a service provider (for example, a telephone company), then you are installing a WAN. One technology, *Asynchronous Transfer Mode* (ATM), which is being introduced at the time of writing, can be used as both a LAN and a WAN. If it becomes as dominant as its supporters claim it will, the distinction between LAN and WAN may disappear.

The channels that connect the computers in a WAN have already been described in section 5.3, but those used to form LANs are quite distinct. Most LANs are connected using traditional copper cables (either twisted-pair or coaxial) and the channel characteristics are tuned for the distances and data rates involved. One major distinction between types of LAN is how the computer systems to be networked are logically connected together. There are many possible topologies that could be used but the four most commonly employed are the star, the tree, the bus and the ring, as shown in Figure 5.5.1. These are how the logical interconnections are configured but

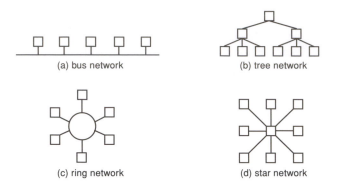

Figure 5.5.1 *LAN topologies*

the physical interconnections may be rather different, for reasons of network reliability or simple logistics. For example, it is very common for a ring-based network to be wired as a star so that any problems with a particular section of the ring can be isolated without a massive rewiring exercise.

Tree topologies were common in the early days of LANs but are hardly used now. Star topologies, with a switch at the centre, are now becoming widespread, not as individual LANs but as a means of interconnecting several LANs. However, the dominant network topologies are those based on a bus and on a ring.

Bus-based local area networks have the characteristic that all the nodes are interconnected by a single physical medium. Consequently there has to be some mechanism to determine which computer can use the communications channel at a given time. This is achieved by a method similar to polite conversation or listen before talk, known as *Carrier-Sense, Multiple Access with Collision Detection* (CSMA/CD), in the most common implementation of a bus-based network, the *Ethernet*. In fact, Ethernet is a proprietary name but is in widespread use to describe LANs implemented to the IEEE 802.3 standard. Unfortunately, CSMA/CD is non-deterministic; that is, you cannot work out a maximum time taken to gain control of the bus from when a station first tries. This (coupled with the associated protocol overheads, see below) makes the usual Information Theory assumptions about data transfer rates rather difficult to apply.

As every node on a bus-based network can communicate with any other node, the electrical characteristics of the channel could vary enormously, depending on the distance between the sender and receiver. For example, a signal transmitted from a point 10 metres away will suffer much less from attenuation and noise than one from 1000 metres away. If the transmitter and receiver are optimised for the former, in terms of the signal strength, they will not be very good at performing the latter connection and vice versa. In order to make the design of the transmitter and receiver possible, limits have to be placed on the maximum and minimum distances between nodes. However, if repeaters or amplifiers are used the maximum distance restriction can be raised, the new limit being fixed by the medium access method.

Bus-based LANs can operate at baseband, that is, the digital signal is applied directly to the transmission channel, or at broadband, with the digital signal modulated on to a high-frequency carrier wave. The advantage of the latter is that it can utilise cable TV technology, such as cheap coaxial cables and amplifiers, to keep the network costs down. However, each node has to have a modem, making those costs higher. Broadband bus-based LANs are dying out now, because they have no overall advantages to compensate for their extra complexity.

Baseband bus-based LANs usually use a line code, such as Bi-phase-L or Manchester (see Chapter 4), to ensure good clock recovery at the receiver. If they use coaxial cables the extra bandwidth, for a given data transmission rate, needed by these line codes is readily available, so rates of 10 Mbps are common. Newer standards, such as 100BaseT (100 Mbps, baseband, twisted-pair), use higher-quality transmitters and receivers so are able to achieve high data rates over twisted pairs, hence reducing the overall wiring costs. In addition, the use of twisted pair wiring allows a logical bus to be installed as a physical star. This gets over one major disadvantage of bus-based LANs, namely their vulnerability to single cable or node failures.

A ring-based network differs in that the interface between the computer and the channel must be active, that is, must have a receiver/transmitter pair. Hence flow of information around the ring is unidirectional. The channel is shared, as with bus-based LANs, but CSMA/CD is not used to gain control of the channel. Commonly a special message, known as a *Token*, is passed from computer to computer. Only the computer holding the token can pass its data over the LAN. Once a computer has gained access to the ring, it must put its message on to the ring and also remove it when it returns. As with any shared medium network, every computer must read the first part of each packet that is going around the ring, in order to determine whether the packet is destined for it.

Most rings operate at baseband, and a wide variety of media can be used. The most common standard for local, ring-based, networks specifies twisted pair wiring and allow data rates up to 16 Mbps. As each node has an active interface there is no need for added repeaters. However, in order to make sure that each node does not suffer an unreasonable delay in getting access to the ring there is a limit to the total number of computers in each ring.

All LANs use CRC codes for error detection, with correction performed by the retransmission of any data received in error. Whatever method is used to implement a computer network, extra information must always be sent along with data being transferred, to help the computer operating system.

Each computer must be able to understand the others in the network, so they must all use the same set of rules for communications. These rules, or sets of rules, are called *protocols*. Unfortunately, there is not one set of protocols that every computer manufacturer can agree on, but there are some standards that most now implement. For example, any computer connected to the Internet must use so-called *TCP/IP* protocols (TCP = Transmission Control Protocol, IP means Internet Protocol). The extra bits used by protocols, including such things as the CRC check bits, reduce the throughput for any LAN. In other words, the actual rates of data transfer that can be achieved are always lower than those indicated by the raw data rate. For example, an Ethernet operating at 10 Mbps will commonly achieve a throughput of only about 3 Mbps (or 30 per cent efficiency). The CSMA/CD channel access algorithm is a major limiting factor here. Token passing rings, which are deterministic, can achieve much higher levels of efficiency, up to 90 or 95 per cent.

5.6   Examples for Solution

1. A system is required for transmitting hand-written diagrams down a standard 3 kHz telephone channel in real time. A proposed system employs an arrangement whereby the position of a pen, moving over a square pad of paper, is monitored in two perpendicular directions by analogue displacement transducers attached to the pen. The outputs of the transducers are digitised and transmitted down the channel as a sequence of binary pulses.

   Discuss the feasibility of the system from the standpoint of Information Theory. Pay particular attention to the resolution of the transmitted diagrams and the speed at which they may be drawn, giving examples of possible resolutions and speeds.

   Compare the applications and advantages of this system with a standard facsimile system.

2. Compare the systems (i), (ii) and (iii) below in terms of the time taken for the transmission of one page of English text along a standard 3 kHz telephone line. Assume that the page measures 30 cm × 20 cm and contains about 300 words.

   (i)   An analogue facsimile system employing double-sideband amplitude-modulation of a 1.5 kHz carrier, with vertical resolution of 0.5 line/mm.

   (ii) An automatic text reader, which identifies each letter in the text and converts it to a five-digit binary code suitable for double-sideband amplitude-modulation as above. (The reader/coder may be assumed to operate instantaneously.)

   (iii) A person reading the page into the telephone at a normal rate.

   Comment on the advantages and disadvantages of the three systems. Compare each of these systems with a hypothetical ideal channel of the same bandwidth and signal-power-to-noise ratio of 30 dB.

3. (a) State an expression for the theoretical capacity C of an information channel in terms of the bandwidth and signal-to-noise ratio, and give a simple interpretation by considering the information per sample in a waveform perturbed by noise. Explain what is meant by the rate R of an information source and how R and C are related by Shannon's Ideal Communication Theorem.

   (b) For each of the two systems A and B described below:
      (i) Discuss the bandwidths and signal-to-noise ratios of the systems, stating and justifying what you consider to be reasonable values. Hence deduce both the theoretical and practical capacities, and explain why the two capacities differ.
      (ii) Discuss the requirements for error protection and describe the coding systems used.

   System A: A facsimile system using a standard telephone channel. You may assume that one A4 page is scanned with a resolution of 1188 × 1728 pixels.

   System B: A Teletext system on a standard television channel. You may assume that one page comprises 24 lines of 40 characters, and is transmitted at the rate of four lines per frame, each frame taking 52 μs.

4. (a) Discuss the principles of error detection and correction using block codes in binary communication systems.

   (b) Discuss the error detection and correction systems in
      (i) teletext systems and
      (ii) compact disc players.

   In each case explain the practical requirements, outline the method used and compare its effectiveness with the levels given by Shannon's Ideal Communication Theorem.

   Note: detailed accounts of the coding systems are not required.

Appendix 1: Probability Theory

A1.1 Basic Probability Theory

We have seen from Chapter 1 that the idea of probability is absolutely basic in Information Theory. We will develop some of the important rules and results of probability theory in this section. In the more mathematical sections of this book a fairly intuitive approach has been adopted, based on an understanding of the processes involved, rather than a strictly rigorous mathematical presentation. Mathematics will be viewed as a tool for the efficient representation and analysis of ideas, rather than a study for its own sake. We will avoid formal proofs except where such proofs are instructive or add to our understanding of the subject.

When we think of the probability of occurrence of an event, such as say the letter 'A' in English text, we mean the relative number of occurrences of 'A' in a long sample of text, so $p(A)$ is the number of occurrences of A divided by the total number of letters in the sample of text. Strictly $p(A)$ is the limit of this ratio as the number of letters in the sample is made indefinitely large. In general, the probability of occurrence of any letter will be given by:

$$p(I) = \lim_{N \to \infty} \frac{N_i}{N}$$

where N_i is the number of occurrences of letter I in a text of N letters. Clearly, the sum of the probabilities of all possible letters must be unity; so

$$\sum_{i=A}^{i=Z} p(i) = 1$$

General Laws

There are two important rules of probability, the sum rule and the product rule. The sum rule applies to events that are mutually exclusive, that is to events that cannot happen at the same time. A single throw of a dice must result in a number from one to six, so the events (one, two, six etc.) are mutually exclusive.

The sum rule simply says that the probability of occurrence of one of a selected set of events is the sum of the probabilities of those events. For example, the probability of throwing a two or a three in one throw of a dice is $1/6 + 1/6 = 1/3$, since the probability of any number is $1/6$. Similarly the probability of throwing one, two, three, four, five or six is clearly 1/6 + 1/6 + 1/6 + 1/6 + 1/6 + 1/6 = 1. Formally:

$$p(A \text{ or } B \text{ or } C \text{ or} \ldots) = p(A) + p(B) + p(C) + \ldots$$

The product rule applies to independent events that occur simultaneously, and states that the probability of a set of such events is given by the product of the individual probabilities. For example, if two dice are thrown together the probability of each being a six is $1/6 \times 1/6 = 1/36$. Formally:

$$p(A \text{ and } B \text{ and } C \text{ and} \ldots) = p(A) \times p(B) \times p(C) \ldots$$

Conditional Probability

Unfortunately many events in the real world are not completely independent, but depend on other events. In a sequence of letters of the alphabet, the probability of say a U after a Q is very high whereas that of a Q after a U is quite low. Such events are known as conditionally or partially dependent events, and the probabilities that describe them as conditional probabilities.

For two conditionally dependent events, A and B, the probability of A and B occurring together is given by:

$$p(A \text{ and } B) = p(A) \times p(B/A) = p(B) \times p(A/B)$$

where $p(A/B)$ means the probability of A given that B has already occurred. The product rule is really a special case of this relation, since if A and B were independent then $p(B/A) = p(B)$ and $p(A \text{ and } B) = p(A) \times p(B)$. Conversely, if A and B were totally dependent then $p(B/A) = 1$ and $p(A \text{ and } B) = p(A)$.

EXAMPLE A1.1

Suppose that the probability of the weather being cold is 0.3, that of it being cloudy is 0.4, and that of it being both cold and cloudy at the same time is 0.2. What is the probability, given that a day is cloudy, of it being cold?

Now $p(\text{cold and cloudy}) = p(\text{cloudy}) \times p(\text{cold/cloudy})$, therefore $p(\text{cold/cloudy}) = 0.2/0.4 = 0.5$. (We can also evaluate $p(\text{cloudy/cold}) = 2/3$.)

Application to Sequences of Letters

When applying these ideas to sequences of letters the same formulae can be used, except that the order of letters is now important. The various letters in a sequence may be conditionally dependent, but the probability of a sequence, say A B, is not necessarily the same as that of B A; in other words $p(A \text{ B}) = p(A) \times p(B/A)$ and $p(B, A) = p(B) \times p(A/B)$ but $p(A, B)$ may not be equal to $p(B, A)$, unlike the case of simultaneous events considered above. $p(A)$ and $p(B)$ will be referred to as 'single' probabilities, $p(A, B)$ and $p(B, A)$ as 'joint' probabilities, and $p(A/B)$ and $p(B/A)$ as conditional probabilities.

The various relationships that exist in such a sequence are conveniently illustrated by an example.

EXAMPLE A1.2

A simple language consists of only two symbols, A and B, produced in a long continuous sequence. Find the single, joint and conditional probabilities of A and B, assuming that the values found from the limited sequence below are typical of a very long sequence (assume that the 21st letter is A in order to have 20 pairs of symbols).

AABBBAAAABBAAABBBAAA

$p(A) = 12/20$, $p(B) = 8/20$ by simply counting As and Bs.
$p(AA) = 9/20$, $p(BB) = 5/20$, $p(AB) = 3/20$, $p(BA) = 3/20$ by counting pairs.
$p(A/B) = 3/8$, $p(B/B) = 5/8$ by counting the number of times an A or B comes after a B.
$p(A/A) = 9/12$, $p(B/A) = 3/12$ by counting As or Bs after an A.

It can be seen that $p(A, B) = p(A) \times p(B/A)$, etc.; indeed, we could have evaluated the conditional probabilities this way instead of counting. It is important to note that the sum of the single probabilities $p(A)$, $p(B)$ is unity, as is that of the joint probabilities $p(AA)$, $p(AB)$, etc. Also, the two groups of conditional probabilities ($p(A/B)$, $p(B/B)$ and $p(A/A)$, $p(B/A)$) both sum to unity. These relations are summarised below, i representing the first symbol in a pair and j the second:

(i) $\sum_i p(i) = \sum_i p(j) = 1$, for example $p(A) + p(B) = 1$

(ii) $\sum_i \sum_j p(i, j) = 1$, for example $p(AA) + p(AB) + p(BA) + p(BB) = 1$

(iii) $\sum_i p(j/i) = 1$, for example $p(A/A) + p(B/A) = 1$.

The single probabilities can be obtained from the joint probabilities from:

(iv) $p(i) = \sum_j p(i, j)$ or $p(j) = \sum_i p(i, j)$, by summing the joint probabilities over the first or second symbols; for example $p(A) = p(AB) + p(AA)$ (summed over second symbols), or $p(B) = p(AB) + p(BB)$ (summed over first symbols). Since $p(A)$ is also given by $p(BA) + p(AA)$ (summed over first symbols) it follows that $p(AB) = p(BA)$ in this case. This is true only for binary sequences and is clearly untrue for the alphabet since $p(QU) \neq p(UQ)$ etc.

A1.2 Probability Distributions

A1.2.1 Discrete Probability Distributions

The Binomial Distribution

The probabilities in a process with two outcomes, such as in tossing a coin, may be described by the Binomial Probability Distribution. Other processes, such as in throwing a dice, may be similarly treated by considering one outcome (say throwing a six) to represent 'success' and all other outcomes to represent 'failure'. Writing p (for success) and $\bar{p}$ (for failure), the probability of r successes in n trials is given by the $(r + 1)$th term of the binomial expansion of $(\bar{p} + p)^n$, that is

$$p(r) = \frac{n!}{(n - r)!r!} \, p^r \bar{p}^{(n-r)}$$

EXAMPLE A1.3

A dice is thrown four times. Find the probabilities of throwing 0, 1, 2, 3 and 4 sixes.
   In this case $p = 1/6$ and $\bar{p} = 5/6$.

$$(\bar{p} + p)^4 = \overline{p^4} \ + \ 4\overline{p^3}p \ + \ 6\overline{p^2}p^2 \ + \ 4\bar{p}p^3 \ + \ p^4$$

r	0	1	2	3	4
$p(r)$	0.4823	0.3858	0.1157	0.0154	0.0008

It is obvious that the probability of getting four sixes is $(1/6)^4$ and of getting no sixes is $(5/6)^4$, but the other probabilities are not intuitively obvious though they are easily found from the formula. Since $(\bar{p} + p) = 1$, the sum of the probabilities on the right-hand side must be unity.
   A plot of $p(r)$ versus r is called a probability distribution, and Figure A1.1 is the distribution for the above example.

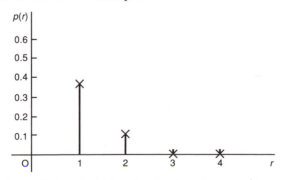

Figure A1.1 *Binomial probability distribution for $p = 1/6$, $n = 4$*

The *mean value* $\bar{r}$ of the distribution is the average number of successes. Clearly

$$\bar{r} = \Sigma \, r \, p(r) = np$$

The *mean square value* $\overline{r^2}$ can be shown to be

$$\overline{r^2} = \Sigma r^2 \, p(r) = (np)^2 + n\bar{p}p$$

The mean square value does not have much significance in discrete distributions, though it is very important in the continuous distributions considered later.

The *variance* σ^2 is a measure of the deviation from the mean, or the width of the distribution, It is given by

$$\sigma^2 = \Sigma(r - \bar{r})^2 \, p(r) = n\bar{p}p$$

and can be shown to be equal to $(\overline{r^2} - (\bar{r})^2)$.

The square root of the variance, σ, is known as the *standard deviation*.

EXAMPLE A1.4

A random sample of 900 persons is asked if they will vote for a certain party in a general election. Find the mean and variance of the distribution, assuming that over the whole population exactly one-third of the people would answer 'yes' and that there are no 'don't knows'. Comment on the result.

The mean value $\bar{r} = np = 900 \times 1/3 = 300$. Similarly $\sigma^2 = 900 \times \frac{1}{3} \times \frac{2}{3} = 200$.

The standard deviation is therefore $\sqrt{200}$ or about 14, and this is the order of the error to be expected statistically between the 'true' value (300) and the result given by the poll. Strictly, if many such polls were carried out the mean would be in the range 300 ± 14 for about 68 per cent of all the polls. Expressed as a percentage of the mean value, the error is a little under 5 per cent. This tends to undermine one's confidence in opinion polls (if one had any), where a swing of 5 per cent is often hailed as a major event.

The Poisson Distribution

It can be shown that if p is very small and n very large, the expression for $p(r)$ can be re-written in the form

$$p(r) = \frac{\exp(-(\bar{r}))(\bar{r})^r}{r!}, \quad \text{depending on } r \text{ only}$$

[It is a little surprising that the exponential (exp) should suddenly appear, but the proof involves replacing the factorials by a power series approximation and comparing the resulting series with that for $\exp(-(\bar{r}))$.]

The most relevant property of the Poisson distribution is that, since p is very small, $\bar{p}$ is very nearly unity so the variance $\sigma^2 = np = \bar{r}$; that is, the variance is equal to the mean value.

The Poisson distribution applies to many physical phenomena, such as radioactive decay, in which n (the number of atoms) is very large but p (the probability of a given atom decaying in a given time) is very low. In particular, it applies to Shot noise in diodes, since there are many electrons with a small probability of crossing the potential barrier.

EXAMPLE A1.5

Over a given period there are 10^6 cars on the roads, and the probability* of a given car having an accident is 10^{-4}. Find the mean number of accidents and the variance.

$$\bar{r} = np = 10^6 \times 10^{-4} = 100$$

$$\sigma^2 = \bar{r} = 100 \text{ so the standard deviation is } 10$$

An increase of 10 per cent in the accident rate over a given period is apparently not necessarily indicative of any real effect, though such figures usually excite many media commentators and politicians.

The Gaussian Distribution

If the number of trials n becomes very large, without p necessarily being small, $p(r)$ can be written in the form

$$p(r) = \frac{1}{\sigma\sqrt{2\pi}} \exp\left(-\frac{(r - \bar{r})^2}{2\sigma^2}\right)$$

depending only on the variance σ^2.

One of the first practical applications of the Gaussian distribution was that it was found to apply to the chest sizes of London guardsmen. It was then found to apply to many other physical characteristics and was soon thought to have some divine significance, though this view has declined somewhat in recent years. It was also found to apply to experimental errors, for example in the repeated measurement of some supposedly fixed quantity.

The explanation of its application to these two apparently different processes is that a given physical characteristic or experimental measurement can be thought of as being due to many small effects (for example, many different genetic factors or many small sources of error). The particular observed value is a particular selection from a binomial distribution (say r factors producing a tall person and $(n - r)$ tending to produce a short one), the Gaussian being a special case of the binomial for large n.

* This rather low value is for male drivers over 45 only.

A similar argument applies to Johnson noise. There are many electrons moving in different directions and the observed noise at an instant depends on how many are moving one way and how many the other way (it also depends on their velocities, of course). The amplitude of Johnson noise follows a Gaussian distribution, as expected.

Although we have introduced the Gaussian distribution as a discrete distribution, it occurs frequently as a continuous distribution. Such distributions occur in many practical situations and will now be considered.

A1.2.2 Continuous Probability Distributions

Probability Density

Suppose we sample a waveform such as that of Figure A1.2(a) with a digital voltmeter, taking a reading with a resolution δx of say, 0.1 V in a time short compared with the rate of change of the waveform. We could produce a histogram giving the number of readings n_x versus range x as in Figure A1.2(b). If we plot n_x/n, where n is the total number of readings, and let n become very large, our ordinate will represent the probability of getting a reading in a given range. The plot will become more symmetrical as n increases, but still jagged to an extent depending on the range interval. Of course, if we reduce the interval δx to 0.01 V or 0.001 V the plot will become more smooth, but will tend to disappear since n_x/n becomes very small. To avoid this, we can divide n_x/n by the interval δx, and let n tend to infinity and δx tend to zero. This process is shown in Figure A1.2(b).

The resulting smooth curve is a plot of probability density $p(x)$ versus value x. The definition of $p(x)$ is

$$p(x) = \lim_{\substack{n \to \infty \\ \delta x \to 0}} \frac{n_x}{n} \frac{1}{\delta x}$$

Clearly $p(x)\, \delta x$ is the probability of a reading being in a small range δx, and the probability of a reading between two values x_1 and x_2 is

$$p(x_1 < x < x_2) = \int_{x_1}^{x_2} p(x)\mathrm{d}x$$

Since any reading must have a value between $x = -\infty$ and $x = +\infty$ then

$$\int_{-\infty}^{+\infty} p(x)\, \mathrm{d}x = 1$$

so the area under the curve is unity.

Similar expressions to those above for the mean, mean square and variance apply. Referring to Figure A1.3 on page 192:

$$\bar{x} = \int_{-\infty}^{+\infty} x\, p(x)\, \mathrm{d}x$$

$$\overline{x^2} = \int_{-\infty}^{+\infty} x^2\, p(x)\, \mathrm{d}x$$

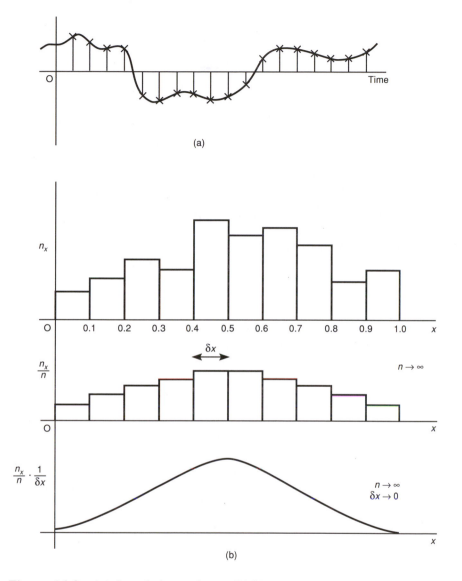

Figure A1.2 *(a) Sampled waveform; (b) histograms and probability density distribution*

$$\sigma^2 = \int_{-\infty}^{+\infty} (x - \bar{x})^2 \, p(x) \, \mathrm{d}x = \overline{x^2} - (\bar{x})^2$$

The mode is the value of x corresponding to maximum $p(x)$ and the median is the value of x dividing the plot into equal areas. The reason that we have to work in terms of probability density when dealing with continuous processes is that the probability of getting an exact value (for example, a

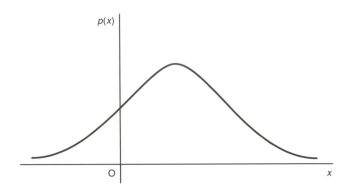

Figure A1.3 *Probability density distribution*

reading of 1.100000001 V) tends to zero. Similarly, the probability of a man having a height of exactly 1.888888 m is very small. We therefore have to use the notion of probability density, and the probability of getting a value in a given range is given by the integral of $p(x)$ over this range, as above.

An expression for the probability density of a continuous waveform can be derived by noting that a waveform spends less time near a given value the greater its rate of change. It can be shown that

$$p(x) = \frac{1}{\text{slope}} \frac{1}{\text{period}}$$

though this has to be used with some care.

EXAMPLE A1.6

Deduce the probability density for the sawtooth waveform of Figure A1.4(a).
    Considering the waveform from 0 to T_0, the slope is A/T_0 so

$$p(x) = \frac{T_0}{A} \frac{1}{T_0} = \frac{1}{A}$$

The maximum value of amplitude is A, so $p(x)$ has the form shown in Figure A1.4(b). This could be deduced intuitively in this case since the waveform moves smoothly through all values between 0 and A. This is known as a uniform distribution; triangular waves have the same distribution.

Important Continuous Distributions

(a) *The Gaussian distribution.* A continuous Gaussian distribution is shown in Figure A1.5.

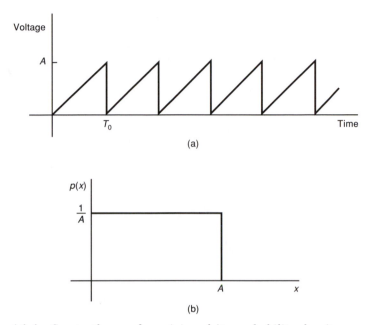

Figure A1.4 *Sawtooth waveform (a) and its probability density distribution (b)*

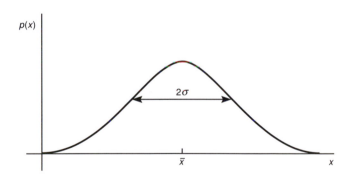

Figure A1.5 *Gaussian distribution*

The probability density $p(x)$ has the same form as for the discrete distribution and is

$$p(x) = \frac{1}{\sigma \sqrt{2\pi}} \, \exp\left(-\frac{(x - \bar{x})^2}{2\sigma^2}\right)$$

The distribution is bell-shaped and symmetrical about its mean value $\bar{x}$. The width between the points of inflexion is 2σ. Since the amplitude of random noise has this distribution it is important to be able to calculate the probability that the noise exceeds a given value. Random noise has a zero mean value, so $\bar{x} = 0$, hence $\overline{x^2} = \sigma^2$.

The probability that a given sample x exceeds a fixed value x_0 is given by

$$p(x > x_0) = \int_{x_0}^{\infty} p(x)\,dx = \int_{x_0}^{\infty} \frac{1}{\sigma\sqrt{2\pi}}\, \exp\left(\frac{-x^2}{2\sigma^2}\right) dx$$

and is represented by the shaded area in Figure A1.6.

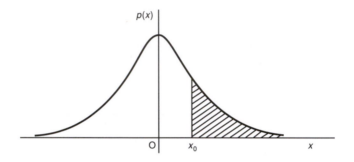

Figure A1.6 *Probability of exceeding a given level for a Gaussian distribution*

Putting $t^2 = x^2/\sigma^2$ so $dt = dx/\sigma$:

$$p(x > x_0) = \frac{1}{\sqrt{2\pi}} \int_{t_0}^{\infty} \exp\left(\frac{-t^2}{2}\right) dt \qquad \text{where } t_0 = x_0/\sigma$$

In communication applications this integral is called the Q function, defined as:

$$Q(x) = \frac{1}{\sqrt{2\pi}} \int_{x}^{\infty} \exp\left(\frac{-t^2}{2}\right) dt$$

The integral cannot be evaluated in closed form (that is, in a simple expression containing functions of t) and numerical methods have to be used. Values are available in standard statistical tables (for example, Howatson *et al.*, 1972) usually in the form

$$\frac{1}{\sqrt{2\pi}} \int_{-\infty}^{x} \exp\left(\frac{-t^2}{2}\right) dt, \quad \text{that is} \quad 1 - Q(x)$$

The proportions of samples falling into given ranges, evaluated from the above formula, are shown in Table A1.1.

When discussing the properties of noise one usually considers the range

Table A1.1 *Proportions of samples in given ranges*

Range	Proportion within range	Proportion outside range
$\pm \sigma$	0.68	0.32
$\pm 2\sigma$	0.95	0.05
$\pm 3\sigma$	0.997	0.003
$\pm 1.96\ \sigma$	95%	5%
$\pm 2.58\ \sigma$	99%	1%
$\pm 3.29\ \sigma$	99.9%	0.1%

as an integral number times σ, whereas in specifying the accuracy of an experimental result one quotes *confidence limits,* being the range required such that, say, 95 per cent of values are within it.

EXAMPLE A1.7

A random telegraph signal has two levels of ± 1 V and is perturbed by random noise of mean square voltage 0.1 V. Find the error rate, assuming that an instantaneous decision is made at the centre of each received pulse with decision level zero.

An error will occur if the signal is positive and the noise less than -1 V or if the signal is negative and the noise greater than 1 V. Therefore

$$p(\text{error}) = p(1) \int_{-\infty}^{-1} p(x)\ \mathrm{d}x + p(0) \int_{1}^{\infty} p(x)\ \mathrm{d}x$$

$$= (p(0) + p(1)) \int_{1}^{\infty} p(x)\ \mathrm{d}x$$

$$= \int_{1}^{\infty} p(x)\ \mathrm{d}x$$

Proceeding as above, putting $t = x/\sigma$ where $x = 1$ and $\sigma = 0.316$ (since $\sigma^2 = 0.1$), we have $t = 1/(0.316) = 3.16$. Therefore

$$p(\text{error}) = \frac{1}{\sqrt{2\pi}} \int_{3.16}^{\infty} \exp\left(\frac{-t^2}{2}\right) \mathrm{d}t = Q(3.16)$$

Using statistical tables, which usually provide $1 - Q(x)$ as explained above, we find $p(\text{error}) = 8 \times 10^{-4}$. (Note that $p(\text{error})$ is independent of $p(0)$ and $p(1)$, this being a binary symmetric system.)

In this example the root mean square signal-to-noise ratio $x/\sigma = 3.16$ is fairly small, though the error rate is surprisingly low. The error rate falls very rapidly with increasing signal-to-noise ratio and is sufficiently low for most practical applications (say 10^{-5} or less) for signal-to-noise ratios of 10 or more.

For a bipolar binary signal, such as that in the above example where the levels are $\pm x$ the error probability is simply $Q(x/\sigma)$. In practice, the signals

used are more often unipolar, for example with the two levels zero and 5 V
and with the decision level set at 2.5 V. The calculation of error rate is
exactly the same and with levels of 0 and x and a decision level of $x/2$ the
general result is

$$p(\text{error}) = \frac{1}{\sqrt{2\pi}} \int_{x/2\sigma}^{\infty} \exp\left(\frac{-t^2}{2}\right) dt = Q(x/2\sigma)$$

In communication systems it is the power-to-noise ratio that is most important and the results are expressed as $Q(P/N)^{1/2}$ and $Q(P/2N)^{1/2}$ for bipolar
and unipolar binary signals respectively (for the unipolar case the signal
power $P = x^2/2$ so $x/(2\sigma) = (P/2N)^{1/2}$).

(b) *The Rayleigh distribution.* When random noise is fed to a bandpass
filter such that the ratio of centre frequency to bandwidth is greater than
about 10, the resulting noise is known as narrow-band noise and has the
appearance of an amplitude-modulated sine wave as shown in Figure A1.7(a).
The noise appears to have a constant frequency but its amplitude fluctuates.
The instantaneous amplitude is known as the envelope, denoted by the dashed
lines. It can be shown that the envelope follows a Rayleigh distribution,
given by

$$p(x) = \frac{x}{\sigma^2} \exp\left(\frac{-x^2}{2\sigma^2}\right)$$

which is shown in Figure A1.7(b).
   The distribution is similar to the Gaussian, but is not symmetrical (since
the envelope cannot be less than zero but has no upper limit). It can be
shown to be formed by the product of two Gaussians. The distribution is
important in communication systems, such as amplitude modulation, in which
the envelope of a sine wave carries the information, since noise then per-
turbs the envelope.

(c) *The uniform distribution.* It can be shown that the phase distribution
of narrow-band noise is uniform, as shown in Figure A1.7(c). An instantaneous measurement of phase (by comparison with a fixed sine wave) will
give any value between $\pm\pi$ with equal probability.

A1.3 Time and Ensemble Averages

The averages that we have dealt with above are known as ensemble averages, as opposed to the time averages that we considered in Chapter 6. Using
the previous notation in each case and considering a continuous random
function:

$$\textit{Ensemble average} \quad \bar{x} = \int_{-\infty}^{+\infty} x\, p(x)\, dx \text{ etc.}$$

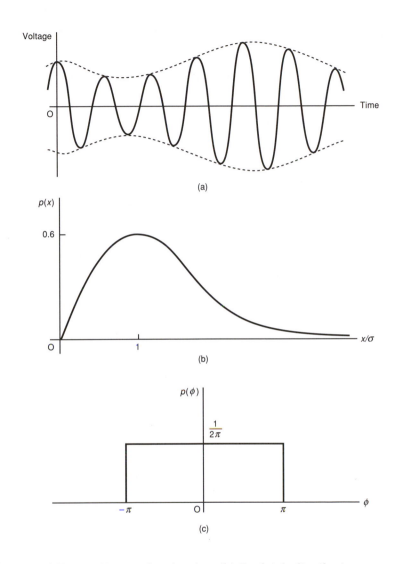

Figure A1.7 *(a) Narrow-band noise, (b) Rayleigh distribution, (c) uniform distribution*

Time average $$\bar{f} \;=\; \lim_{T \to \infty} \frac{1}{T} \int_{-T/2}^{T/2} f(t)\, dt \text{ etc.}$$

Both averages should clearly give the same result if applied to the same signal, and the relation between them can be seen by considering the Johnson noise produced by an infinite set of nominally identical resistors, as shown in Figure A1.8.

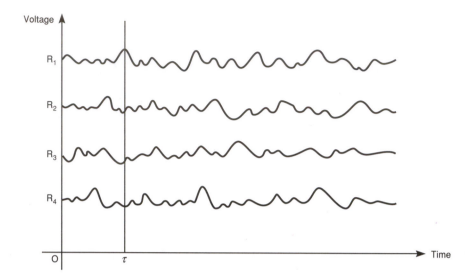

Figure A1.8 *Noise waveforms for a set of 'identical' resistors*

The ensemble average is effectively obtained from the complete set of resistors at any time τ; the voltages produce a probability density distribution from which $\bar{x}$ is evaluated. The time average is obtained from any one resistor in the set, averaging over all time. Clearly, the two values will be the same.

EXAMPLE A1.8

Find the mean, mean square and variance for (i) a sine wave, (ii) random noise and (iii) a triangle wave, using time averages or ensemble averages as appropriate.

(i) *Sine wave.* Since the equation for a sine wave is $x = a \sin \omega_0 t$ it is easier to use a time average. We obtain $\bar{x} = 0$, $\overline{x^2} = \sigma^2 = a^2/2$.

(ii) *Random noise.* $x(t)$ cannot be predicted in this case, so we have to use an en-semble average:

$$p(x) = \frac{1}{\sigma\sqrt{2\pi}} \exp\left(-\frac{x^2}{2\sigma^2}\right)$$

so $\bar{x} = 0$, $\overline{x^2} = \sigma^2$.

(iii) *Triangle wave.* Using the waveform of Figure A1.9, $x(t)$ is given by

$$x = \frac{A}{T_0/2} t \qquad \text{from 0 to } T_0/2$$

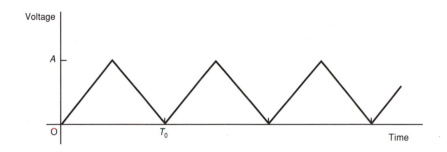

Figure A1.9 *Triangle waveform*

Using time averages

$$\bar{x} = \frac{1}{T_0/2} \int_0^{T_0/2} \frac{A}{T_0/2} t \, dt = \frac{A}{2} \quad \text{and similarly} \quad \overline{x^2} = \frac{A^2}{3}$$

However, we can also use an ensemble average since $p(x) = 1/A$:

$$\bar{x} = \int_0^A \frac{1}{A} x \, dx = \frac{A}{2}, \qquad \overline{x^2} = \frac{A^2}{3}$$

A1.4 Examples for Solution

1. A probability density distribution is given by $p(x) = a \exp(-|x|/b)$, where a and b are constants.
   (i) Find the mean, mean square and variance in terms of b.
   (ii) Noise having this distribution is added to a random binary telegraph signal of amplitude $\pm V$. Find the error probability, assuming the decision level is zero, and sketch as a function of V.
2. Many full-wave rectifier meters are calibrated so that the indicated root-mean-square value is correct for sine wave inputs. Find the ratio between the mean modulus and the root-mean-square value of a sine wave and hence deduce the percentage error when such a meter is used for the measurement of (i) square waves and (ii) Gaussian noise.
   Note: the mean modulus $|x|$ is the mean value ignoring the sign.
3. (a) Find and sketch the probability density amplitude distributions for a square wave, a triangle wave and a sine wave.
   (b) Find the mean square values of these functions, (i) as an ensemble average and (ii) as a time average.

Appendix 2: Electrical Noise

A2.1 Types of Electrical Noise

Random electrical noise occurs in any practical information transmission system. Although its general characteristics such as mean-square value and power spectrum may be known, its essential characteristic is that its value at any instant (amplitude or frequency) is unpredictable. Electrical noise can, of course, be considered to be a signal, and is capable of carrying information; in fact it can be shown that noise-like waveforms have the greatest information capacity. However, in practical systems electrical noise is an unwanted signal, producing errors or changes in amplitude of the wanted signal, and reducing the overall information transfer.

A2.1.1 Johnson Noise

Johnson noise is the most basic type of electrical noise. It was discovered by Johnson (1928) and the expression for its power spectrum deduced by Nyquist (1928). Johnson noise occurs in any device that dissipates power, such as a resistor, but since all devices have some resistance (for example, capacitors, inductors, transistors) its occurrence is universal. It is the electrical equivalent of mechanical Brownian motion, and the defining equations are identical in form.

The Principle of Equipartition of Energy states that the mean kinetic energy of each degree of freedom of a system in thermal equilibrium with its surroundings in $\frac{1}{2} kT$, where k = Boltzmann's constant (1.4×10^{-23} J/K) and T is the absolute temperature. The electrons in a conductor thus have a mean square velocity (in one direction) that is dependent on temperature and given by $\frac{1}{2} mv^2 = \frac{1}{2} kT$. Since an electron is a charged particle, a moving electron constitutes a current and Johnson noise is the net effect, at any instant, of the current pulses due to all the electrons in the conductor.

The process is actually rather more subtle than this. An electron moving through the conductor with a velocity V produces a steady current eV. Other electrons moving in different directions with different velocities similarly produce steady currents, and the net effect of all these currents would average to zero. However, the essential point is that electrons do not flow freely through conductors; they collide with the crystal lattice and in doing so change their velocities. Johnson noise is the net effect of the current pulses due to collisions, as illustrated in Figure A2.1. The mean value averages to zero but the mean-square value does not. In a perfect conductor there would be no collisions and no noise; however, collisions with the crystal lattice

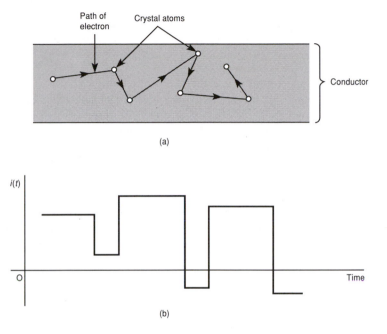

Figure A2.1 *(a) Movement of electrons in crystal lattice. (b) Current pulses corresponding to (a).*

are the mechanism whereby thermal equilibrium is achieved, so a perfect conductor could not attain thermal equilibrium in any case. It can be said that dissipation is the cause of fluctuations, since thermal equilibrium occurs via dissipation and fluctuations result from it.

It can be shown that the power spectrum (in mean-square volts per Hz) for the Johnson noise in a resistor R is given by

$$P_v(f) = 4RkT$$

This is constant up to very high frequencies, and is referred to as *white noise*, since it contains all frequencies, in analogy to white light, which contains all colours.

A2.1.2 Shot Noise

Shot noise is less universal than Johnson noise, but occurs in diodes and hence in transistors, so in practice it is often the main source of noise. It was first studied in thermionic diodes, in which the emission of electrons from the cathode is a random process, so that the number of electrons reaching the anode per second fluctuates about its mean value. The effect occurs similarly in semiconductor diodes, since the passage of an electron across the potential barrier in a p–n diode is again a statistical process, leading to a fluctuation in the current. The expression for the power spectrum (in terms of mean-square current per Hz) is

$$P_i(f) = 2eI$$

where I is the mean current and e the electronic charge (1.6×10^{-19} C).

The spectrum is again constant up to very high frequencies. Unlike Johnson noise, Shot noise depends specifically on the magnitude of e. If electricity were continuous there would be no Shot noise, but the formula for Johnson noise would be unchanged.

A2.1.3   1/f Noise

It is found that most semiconductor devices produce more noise, particularly at low frequencies, than is predicted by Johnson or Shot noise. This is attributed to $1/f$ noise, which has a power spectrum inversely proportional to frequency, given by

$$P_v(f) = K \frac{I^2}{f}$$

in which I is the current in the device and K is a constant that depends on the device. A number of theories have been put forward to account for $1/f$ noise, but there is no satisfactory single quantitative explanation. The effect does not occur in pure metals, but has been observed over many decades of frequency in semiconductors. It is apparently due to fluctuations in the numbers of electrons in the conduction band and 'queueing' theories have been favoured.

The formula is a little disturbing in that it predicts infinite noise at zero frequency. In practice, of course, one is rarely interested in averaging over periods greater than a few seconds, and even a day (10^5 s or 10^{-5} Hz) is far from zero frequency. Interestingly, $1/f$ noise has the property that the noise power in equal frequency ratios (decades, octaves etc.) is the same, unlike white noise where the power is the same in equal intervals. There is therefore five times as much $1/f$ noise power between 1 Hz and 10^{-5} Hz as between 10 Hz and 1 Hz. A power spectrum for the noise from a typical operational amplifier is shown in Figure A2.2.

A2.2   Properties of Electrical Noise

A2.2.1   Time and Frequency Domains

The essential nature of electrical noise is that it is unpredictable in the time domain, its amplitude following a Gaussian distribution; in the frequency domain it is white, that is, its power spectrum is constant.

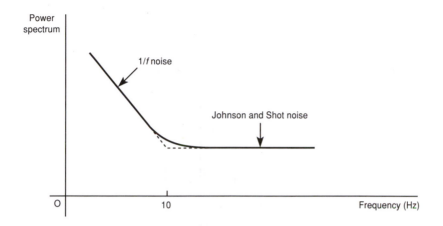

Figure A2.2 *Noise power spectrum of operational amplifier*

Time domain: $p(v) = \dfrac{1}{\sigma\sqrt{2\pi}}\, e^{\frac{-v^2}{2\sigma^2}}$

Frequency domain: $P(f) = $ constant

These two properties are really different ways of saying the same thing; unpredictability in the time domain implies flatness in the frequency domain, and vice versa. If the spectrum was not flat, but fell away at higher frequencies, it would be possible to predict accurately the amplitude a short time ahead. Of course, the spectrum does fall away eventually, otherwise the total power would be infinite, but for practical purposes it is considered to be flat.

Another way of looking at predictability is afforded by the concept of the autocorrelation of a waveform, introduced in Chapter 2.

A2.2.2 Representation of Noise

Figure A2.3 shows the power spectrum of random noise, divided into strips each of width δf.

One way of representing noise mathematically is to replace each strip by a single sinusoidal oscillator whose frequency is that of the centre of the strip and whose amplitude is equivalent to the power in the strip. Thus, the ith strip is represented by an oscillator $a_i \cos(\omega_i t + \varphi_i)$, where $\frac{1}{2} a_i^2 = P_0\, \delta f_i$, φ_i being the (random) phase. The noise waveform can be written as $v_n(t) = \Sigma\, a_i \cos(\omega_i t + \varphi_i)$ and since all the amplitudes a_i are the same

$$v_n(t) = (2P_0\, \delta f_i)^{\frac{1}{2}} \Sigma \cos(\omega_i t + \varphi_i)$$

A more useful representation for narrow-band noise can be obtained from

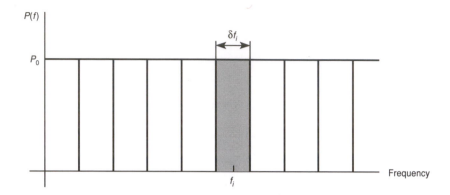

Figure A2.3 *Power spectrum of random noise*

this expression by writing the frequency of the *i*th strip as $f_i = f_0 + i\ \delta f_i$ and expanding $\cos(\omega_i t + \varphi_i)$. The resulting expression can be written as

$$v_n(t) = r(t)\ \cos(\omega_0 t + \varphi(t))$$

where $r(t)$ involves a summation but is in fact just the envelope of the noise discussed above, so $r(t)$ is Rayleigh distributed and $\varphi(t)$ uniform. A further representation deducible from this is

$$v_n(t) = x(t)\ \cos\ \omega_0 t + y(t)\ \sin\ \omega_0 t$$

where $x(t)$ and $y(t)$ again involve summations, but are both Gaussian distributed.

The importance of these two representations is that they involve only a single frequency f_0 (the centre frequency of the filter) so the effect of noise on a system can be found very easily. For example, if noise represented by the second equation is added to a sinusoidal carrier $a\ \cos\ \omega_0 t$, then $x(t)$ produces amplitude errors and $y(t)$ produces phase (or frequency) errors.

A2.2.3 Addition of Random Generators

When two independent sources of noise are added, it is the mean-square values which add, in contrast to two signal sources where the root-mean-square values are added. That is

Noise sources: $\overline{v^2_{sum}} = \overline{v_1^2} + \overline{v_2^2}$

Signal sources: $\sqrt{\overline{v^2_{sum}}} = \sqrt{\overline{v_1^2}} + \sqrt{\overline{v_2^2}}$

The reason for this can be seen by considering the two noise sources of Figure A2.4.

The instantaneous sum $v_{sum}(t) = v_1(t) + v_2(t)$. The mean-square value is

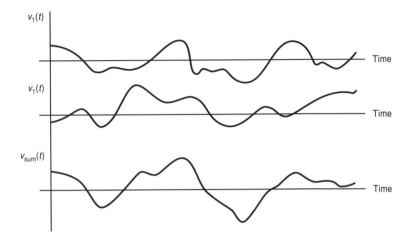

Figure A2.4 *Waveforms of two independent noise sources*

$$\overline{v^2_{sum}} = \overline{v_1^2} + \overline{v_2^2} + \overline{2v_1v_2}$$

Since the noise sources are independent, the cross term $2v_1v_2$ averages to zero. However, for two signals of the same frequency and phase (but different amplitudes, say), $\overline{2v_1v_2}$ would not be zero (the contributions at all times would be positive), so we would find

$$\overline{v^2_{sum}} = ((\overline{v_1^2})^{\frac{1}{2}} + (\overline{v_2^2})^{\frac{1}{2}})^2$$

In other words the root-mean-square values add.

Noise sources are said to be incoherent and signal sources coherent.

EXAMPLE A2.1

A thermionic diode is connected in series with a resistor of 10 kΩ and carries a current of 0.1 mA, as shown in Figure A2.5. Find the noise voltage at the output in a bandwidth of 1 MHz.

The resistor produces Johnson noise $4RkTB$ in bandwidth B. Taking $kT = 4 \times 10^{-21}$ J at room temperature, $\overline{v_J^2} = 4 \times 10^4 \cdot 4 \times 10^{-21} \cdot 10^6 = 1.6 \times 10^{-10}$ V^2.

The diode produces a Shot noise current, which flows through the resistor, so the voltage produced is $\overline{v_s^2} = 2eIBR^2 = 2 \times 1.6 \times 10^{-19} \cdot 10^{-4} \cdot 10^6 \cdot 10^8 = 3.2 \times 10^{-9}$ V^2

The total noise, the sum of the mean-square values, is 3.4×10^{-9} V^2 or about 5.9×10^{-5} V root mean square.

Note: It may appear rather academic to consider a thermionic diode, which is little used in practice now. However, the reason for doing so is to simplify

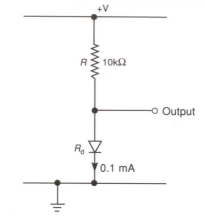

Figure A2.5 *Diode and resistor network for noise analysis*

the calculations since the Johnson and Shot noise can be directly added. A thermionic diode has a very high impedance so that the noise voltages are measured across the 10 kΩ resistor. In contrast, a semiconductor diode has a low impedance (250 Ω at a current of 0.1 mA), so the noise voltage is determined by this small value, and is much smaller (see example A2.4, no. 2).

A2.3   Noise in Cascaded Networks

All electrical networks (such as amplifiers, filters, demodulators) will add noise to any signal that passes through them. This noise will be generated by thermal effects in resistors and by the various semiconductor devices in the network. What is important as far as the communications engineer is concerned is the total effect of this added noise on the signal-to-noise ratio. It is common to define a quantity, called the *Figure of Noise (F)* (sometimes *Noise Figure*), to account for all sources of noise within the network:

$$F = \frac{\text{SNR}_{\text{in}}}{\text{SNR}_{\text{out}}}$$

Note that as noise will always be added by the network, F will always be greater than unity. Consider the network in Figure A2.6.

Signal power at output of network $= G_1 S_{\text{in}}$

Noise power at output of network $= G_1 N_{\text{in}} + N_1$

Noise Figure of network $F_1 = \dfrac{S_{\text{in}}}{N_{\text{in}}} \times \dfrac{G_1 N_{\text{in}} + N_1}{G_{\text{in}} S_{\text{in}}} = \dfrac{G_1 N_{\text{in}} + N_1}{G_1 N_{\text{in}}}$

If the signal passes through more than one network then an overall Figure

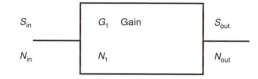

Figure A2.6 *A noisy electrical network*

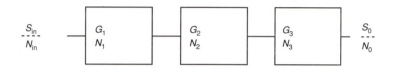

Figure A2.7 *Cascaded noisy electrical networks*

of Noise can be calculated. Consider the following example, for three electrical networks in series, as shown in Figure A2.7:

Signal power at output of system $= G_1 G_2 G_3\, S_{in}$

Noise power at output of system $= G_1 G_2 G_3\, N_{in} + G_2 G_3\, N_1 + G_3\, N_2 + N_3$

Hence, Overall Figure of Noise, $F_o = \dfrac{S_{in}}{N_{in}} \times \dfrac{G_1 G_2 G_3\, N_{in} + G_2 G_3\, N_1 + G_3\, N_2 + N_3}{G_1 G_2 G_3\, S_{in}}$

$$= 1 + \frac{N_1}{G_1\, N_{in}} + \frac{N_2}{G_1 G_2 N_{in}} + \frac{N_3}{G_1 G_2 G_3\, N_{in}}$$

or, using the Figure of Noise for each network:

$$F_o = F_1 + \frac{F_2 - 1}{G_1} + \frac{F_3 - 1}{G_1 G_2}$$

This is an important result because it emphasises the role of the noise performance of the first network in a chain. It is vital that this is made as low noise as possible (perhaps by cooling) in order to lower the overall reduction in SNR suffered as the signal passes through the chain. An example of this is the use of low-noise amplifiers as the first stage in a satellite receiver.

Another common approach is to represent all the sources of noise in a network by an equivalent thermal noise source, giving rise to the concept of an *Equivalent Noise Temperature* (T_e). The result derived above for cascaded noisy networks could be written as

$$T_{eo} = T_{e1} + \frac{T_{e2}}{G_1} + \frac{T_{e3}}{G_1 G_2}$$

where T_{ex} is the equivalent noise temperature for each of the networks in turn.

A2.4 Examples for Solution

1. (i)   Explain why Johnson noise increases with temperature. Does
          it occur at absolute zero?
   (ii) Does Johnson noise depend on the direct current flowing in a resistor?
          Does it occur in semiconductors?
   (iii) Does Shot noise depend on the current, on the temperature or on
          both in a semiconductor diode? Does Shot noise occur in a resistor
          carrying direct current?

2. Repeat example 1 but using a semiconductor diode instead of the thermionic
   diode. What would be the effect on the noise if the current was increased
   by a factor of 10 for (i) the thermionic diode and (ii) the semiconductor
   diode?
   (*Note*: the resistance of a silicon diode is 25 Ω at 1 mA, 250 Ω at 0.1
   mA, etc.)

Appendix 3: Fourier Series

A3.1 Fourier Theory

Fourier Theory uniquely relates the time and frequency domain representations of signals. It is a particularly satisfying theory, being mathematically precise and elegant, and yet having a very clear physical interpretation and application. Like most worthwhile studies its understanding requires a certain amount of initial effort. It has been said that 'Fourier Theory grows on you' but unfortunately this is rather undermined by the reply 'like warts or spots'. Most subjects become easier with use, of course, but this is particularly true of Fourier Theory. It is very difficult to find an examination question that someone well versed in the subject cannot do in less than about ten minutes; indeed, this is probably the best academic reason for not teaching it!

The subject is beloved of mathematicians who, by their nature, are rather more interested in the mathematics than its application. Probably because of this it has not been much liked by engineers, who set great store by the rather similar Laplace Transform yet tend to shun the Fourier Transform. This is a pity, because the former is used mainly as a tool in finding the time domain response of systems but is much less elegant and has a far less obvious physical meaning than the latter.

It is, of course, possible to make some progress in Fourier Theory simply by learning the rules and turning the handle (as is often done with Laplace Transforms), but such an approach is ill advised. Understanding is everything in science. One can teach a dog all sorts of tricks, but as soon as some new situation occurs they are all useless. It is essential to make the effort necessary to understand basic scientific theory; it is well worth the effort and can be very rewarding. Otherwise, unfortunately, one always gets caught out sooner or later.

As explained in Chapter 2 it is perhaps easiest for normal persons (that is, non-mathematicians) to be introduced to Fourier Theory by way of Fourier Series, which apply only to repetitive signals (that is, finite power signals), and this appendix is devoted to this topic.

A3.2 Fourier Series

The essential idea of Fourier Series is that a repetitive function of time $f(t)$ of period T can be represented as a sum of simple waveforms, that is

$$f(t) = a_0 + \sum_{n=1}^{n=\infty} a_n \cos n\omega_0 t + \sum_{n=1}^{n=\infty} b_n \sin n\omega_0 t \qquad (A3.1)$$

209

where $\omega_0 = 2\pi/T$ and a_0, a_n and b_n are constants. The meaning becomes clearer if we combine the second and third terms, giving

$$f(t) = d_0 + \sum_{n=1}^{n=\infty} d_n \cos(n\omega_0 t + \phi_n) \qquad \text{(A3.2)}$$

where $d_n = (a_n^2 + b_n^2)^{\frac{1}{2}}$ and $\phi_n = \tan^{-1}(-b_n / a_n)$.

This equation means that a repetitive function of time is equivalent to the sum of a set of cosine waveforms at angular frequencies ω_0, $2\omega_0$, $3\omega_0$ etc., each with different amplitude d_n and phase ϕ_n, plus a constant level d_0 ($= a_0$). The constituent waveforms are known as harmonics; the zeroth harmonic has zero frequency and amplitude d_0; the first harmonic has frequency $1/T$ ($= \omega_0/2\pi$), amplitude d_1 and phase ϕ_1; the second has frequency $2/T$, amplitude d_2 and phase ϕ_2 etc.

It can be shown (see for example Stuart, 1966) that a_0, a_n and b_n are given by

$$a_0 = \frac{1}{T} \int_{-T/2}^{T/2} f(t) \, dt \qquad \text{(A3.3)}$$

$$a_n = \frac{2}{T} \int_{-T/2}^{T/2} f(t) \cos n\omega_0 t \, dt \quad \text{and} \quad b_n = \frac{2}{T} \int_{-T/2}^{T/2} \sin n\omega_0 t \, dt$$

Equation A3.3 could have been written in terms of sine waves, of course, but cosines are preferable.

EXAMPLE A3.1

Find the coefficients a_0, a_n, b_n for the square wave of Figure A3.1.

$$a_0 = \frac{1}{T} \int_{-T/2}^{T/2} f(t) \, dt = \frac{1}{T} \int_{-T/4}^{T/4} A \, dt = \frac{A}{2}$$

$b_n = 0$ since $f(t)$ is even and the integral is odd.

$$a_n = \frac{2}{T} \int_{-T/4}^{T/4} A \cos n\omega_0 t \, dt = \frac{2A}{n\pi} \sin n \frac{\pi}{2}$$

Thus:

$$f(t) = \frac{A}{2} + \frac{2A}{\pi} \left(\cos\omega_0 t - \frac{1}{3} \cos 3\omega_0 t + \frac{1}{5} \cos 5\omega_0 t \ldots \right)$$

The plot of d_n versus f is known as the amplitude spectrum, and that of ϕ_n versus f is known as the phase spectrum. These are shown in Figure A3.1(b) and (c).

The zeroth harmonic is just the mean value $\bar{f}$. For a square wave all the even harmonics are zero and the odd ones decrease uniformly in amplitude (as $1/n$) and alternate in phase between zero and $-\pi$.

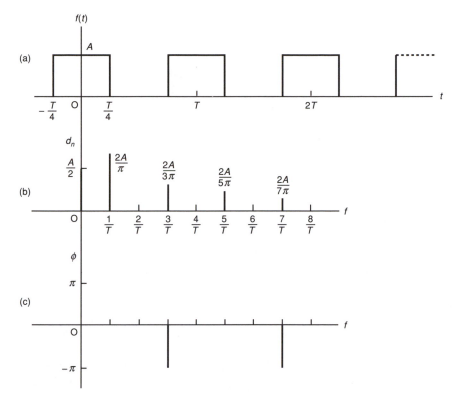

Figure A3.1 *(a) Square wave, (b) amplitude spectrum, (c) phase spectrum*

Equations A3.1 and A3.2 are reversible. The time domain representation $f(t)$ is completely equivalent to its frequency domain representation d_n and ϕ_n and each uniquely determines the other. A repetitive signal can be broken down into its harmonics by feeding it to a suitable narrow-band tuneable filter, and the same waveform can be reconstructed by adding together the appropriate harmonics. There are, however, some practical problems in demonstrating this experimentally. One must measure both the amplitude and phase of the harmonics but narrow-band filters have phase characteristics that change very rapidly with frequency, so that it is difficult to measure the phase accurately. Similarly, in attempting to reconstruct a signal from its (known) harmonics the frequencies have to be exact multiples of $1/T$; even small errors will severely affect the waveform produced. To demonstrate the reconstruction in practice the harmonics have to be derived (by division or multiplication) from the same master oscillator. However, the availability of microcomputers has greatly improved the situation; one can easily generate and add the required waveforms in software, and a convincing demonstration

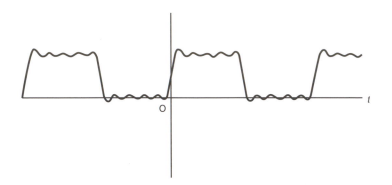

Figure A3.2 *Fourier synthesis of a square wave*

of Fourier synthesis can be given. Figure A3.2 is a print-out of such a synthesis for a square wave, using the first nine harmonics.

A3.2.1 Complex Representation of Fourier Series

The equations of Fourier synthesis (A3.1 and A3.2) and the corresponding equations for the coefficients (A3.3) can hardly be said to be very elegant or compact. Equation A3.1 has three terms on the right-hand side involving the three separate coefficients of equation A3.3. However, a considerable simplification can be achieved by replacing the sines and cosines by cisoids, using the relation $\exp(j\phi) = \cos\phi + j\sin\phi$. The word cisoid comes from $\cos + i\sin$ but most engineers now use j to represent the imaginary number i. Perhaps there was some good reason when this was established, or maybe they wanted to appear different from mathematicians and physicists, both of whom usually use 'i'. After a little manipulation we obtain:

$$f(t) = \sum_{n=-\infty}^{n=\infty} c_n \exp(jn\omega_0 t) \tag{A3.4}$$

$$c_n = \frac{1}{T} \int_{-T/2}^{T/2} f(t)\exp(-jn\omega_0 t)\mathrm{d}t$$

$c_n = \frac{1}{2}(a_n - jb_n)$ is the complex amplitude spectrum and replaces d_n and ϕ_n above. Equations A3.4 are much the most satisfactory and compact statement of the Fourier Series relations, but require a little explanation. $f(t)$ is now represented by a sum of cisoids, over both positive and negative values, and c_n is given by a single integral again involving a cisoid. Figure A3.3 shows the cisoids $\exp(j\omega_0 t)$ and $\exp(-j\omega_0 t)$ plotted on an Argand diagram; $\exp(j\omega_0 t)$ rotates anticlockwise with angular velocity ω_0 tracing out a circle of unit radius, and $\exp(-j\omega_0 t)$ rotates similarly clockwise.

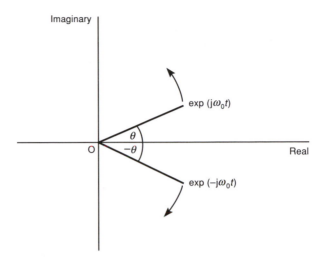

Figure A3.3 *Positive and negative cisoids*

It is well known that

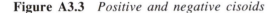

$$\cos \omega_0 t = \tfrac{1}{2} \exp(j\omega_0 t) + \tfrac{1}{2} \exp(- j\omega_0 t)$$

and this is immediately evident from Figure A3.3; if we add the two cisoids, the imaginary components cancel and the real parts add (hence the factor $\tfrac{1}{2}$ above).

Cisoids are very important basic functions in Signal Theory, much more so than simple sine waves and cosine waves. Any signal, whether real, imaginary or complex, can be constructed by adding cisoids, whereas sine and cosine waves are essentially real. However, many signals in information processing have to be expressed in complex form. If a signal has both amplitude and phase it can be expressed by a single complex number $z = x + jy = |z| \exp(j\phi)$, the real modulus $|z|$ ($= (x^2 + y^2)^{1/2}$) and phase ϕ ($= \tan^{-1} y/x$) being both expressed by the complex z. This applies to the response of filters (having amplitude and phase response), to optical systems (transmission amplitude and phase) and to many other systems as well as to the basic Fourier Theory.

People sometimes claim that cisoids are 'abstract' and try to avoid using them. However, this is rather like trying to do something (such as make love) with both hands tied behind one's back; it may present something of a challenge but it considerably limits the possibilities. Most mathematical concepts are rather abstract but cisoids are hardly more so than the notion of a sine wave as the projection on the y-axis of a point moving in a circle!

The meaning of equation A3.4 is that any repetitive function $f(t)$ can be represented as a sum of cisoids, the amplitudes and phases of the various harmonics being given by the modulus and phase of the complex function c_n. Since cisoids may be either positive or negative, the spectrum has both positive and negative sides (n goes from $-\infty$ to $+\infty$). However, for a real function $f(t)$, the positive and negative cisoids for a given value of n combine to give a real cosine waveform, since $\exp(jn\omega_0 t) + \exp(-jn\omega_0 t) = 2 \cos \omega_0 t$ as above. One can think of a real function $f(t)$ as comprising both positive and negative cisoids, but that we are physically unable to distinguish between them.

The equation for c_n can be written

$$c_n = \frac{1}{2}(a_n - jb_n) = \frac{1}{T}\int_{-T/2}^{T/2} f(t)[\cos n\omega_0 t - j \sin n\omega_0 t]\,dt$$

so

$$a_n = \frac{2}{T}\int_{-T/2}^{T/2} f(t) \cos n\omega_0 t\,dt \tag{A3.5a}$$

and

$$b_n = \frac{2}{T}\int_{-T/2}^{T/2} f(t)\sin n\omega_0 t\,dt \tag{A3.5b}$$

agreeing with equation A3.3 (c_0 is identified with $2a_0$). If $f(t)$ is even then $b_n = 0$ (since $\sin n\omega_0 t$ is odd) and similarly if $f(t)$ is odd then $a_n = 0$; in these cases one could use only equation A3.5a or equation A3.5b but the saving in calculation is small and usually it is easier to stick to equation A3.4.

EXAMPLE A3.2

Repeat example A3.1 using the complex representation.

$$c_n = \frac{1}{T}\int_{-T/4}^{T/4} A \exp(-jn\omega_0 t)\,dt = \frac{A}{n\pi} \sin \frac{n\pi}{2}$$

The spectrum is shown in Figure A3.4 ($c_0 = A/2$ since $\lim_{x\to 0}(\sin x/x) = 0$).

Having found c_n we simply have to find the modulus to get the amplitude spectrum and the argument to get the phase. The plots represent the distribution of cisoids, in terms of the (positive or negative) harmonic number n. However, it is usual to label the axis as frequency, so that lines appear at intervals of $1/T$. It should be noted that the plot for the amplitude spectrum does not have an 'out of sequence' value at the origin, unlike that in Figure A3.2, and that the amplitude spectrum is even and the phase spectrum odd.

This is always the case for a real function, because the imaginary parts of the constituent cisoids must cancel and they can do this only if $c(+n) = c(-n)$ and $\phi(+n) = -\phi(-n)$, as can be seen by expanding equations A3.4.

Physically, of course, we measure only positive frequencies, so the measured amplitude spectrum is the sum of the two sides (often thought of as a 'folding over' of the negative side on to the positive) and the measured phase spectrum is (by convention) the positive side.

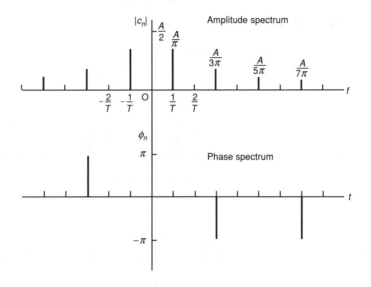

Figure A3.4 *Spectrum of a square wave using complex form of Fourier Series*

A3.2.2   Sequence of Rectangular Pulses

The most instructive example in Fourier Series is an infinite sequence of rectangular pulses, since many important deductions can be made from it. In Figure A3.5(a) the origin has been deliberately chosen at the left-hand edge of a pulse, to give a more interesting phase spectrum (the analysis is slightly easier with the origin in the centre of a pulse).

$$c_n = \frac{1}{T} \int_0^\tau A \, \exp(jn\omega_0 t) dt$$

$$= \frac{A}{T} \frac{-1}{jn\omega_0} [\exp(-jn\omega_0 t]_0^\tau$$

$$= \frac{A}{T} \frac{-1}{jn\omega_0} [\exp(-jn\omega_0 \tau] - 1]$$

$$= \frac{A}{T} \frac{-1}{jn\omega_0} \exp(-jn\omega_0 \tau/2)[\exp(-jn\omega_0 \tau/2) - \exp(+jn\omega_0 \tau/2)]$$

$$= \frac{A}{T} \frac{-1}{jn\omega_0} \exp(-jn\omega_0 \tau/2) - (2j \sin(n\omega_0 \tau/2))$$

$$= \frac{A}{T} \frac{-1}{jn\omega_0} \exp(-jn\omega_0 \tau/2) - \left(2j \frac{\sin(\pi n\tau/T)}{\pi n\tau/T} \frac{\pi n\tau}{T}\right)$$

which simplifies to

$$c_n = \frac{A\tau}{T} \exp(-jn\pi\tau/T)\mathrm{sinc}(n\tau/T)$$

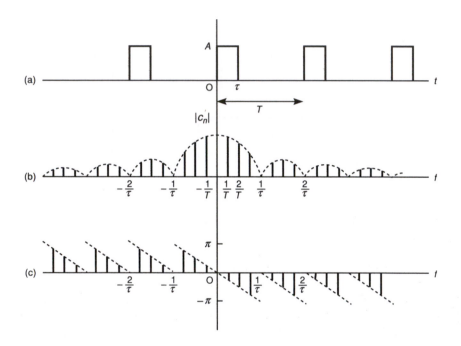

Figure A3.5 (a) Sequence of rectangular pulses, (b) amplitude spectrum, (c) phase spectrum

The *sinc* function is defined by: $\mathrm{sinc}(x) = \dfrac{\sin(\pi x)}{\pi x}$

and is plotted in Figure A3.6. It is very important in Communication Theory and occurs whenever a rectangular shaped signal occurs. It has the interesting property that it is zero for all integral values of x (other than $x = 0$) and that both $\int_{-\infty}^{\infty} \mathrm{sinc}(x)\mathrm{d}x$ and $\int_{-\infty}^{\infty} \mathrm{sinc}^2(x)\,\mathrm{d}x$ evaluate to unity. (These properties would not be so simple had it been defined as $\sin x/x$.)

Evaluating the modulus and phase of c_n we obtain the spectra (b) and (c) in Figure A3.5, which are plotted for a ratio of period to pulse width $T/\tau = 4$.

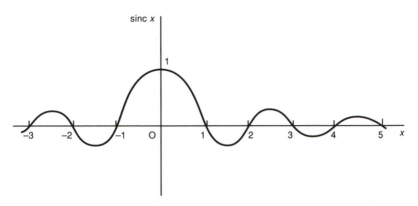

Figure A3.6 *The sinc function*

The harmonics are spaced at intervals of $1/T$, their envelope following the (modulus) of the sinc function. A zero amplitude occurs whenever $\text{sinc}(n\tau/T)$ is integral so with $T/\tau = 4$ the fourth, eighth, twelfth lines etc. are zero; these zeros occur at frequencies $1/\tau$, $2/\tau$, $3/\tau$ etc., directly determined by the pulse width. There are clearly two factors that determine the spectrum: the repetition of the waveform produces lines every $1/T$ Hz and the envelope of the spectrum is determined by the shape of the waveform. The term $\exp(-jn\pi\tau/T)$ is a phase term dependent on the choice of origin and vanishes if the origin is chosen in the centre of a pulse. In general, a shift of origin of θ in time produces a phase term of $\exp(-jn\omega_0\theta)$ in the corresponding spectrum.

A number of useful deductions can be made and are illustrated in Figure A3.7:

(i) For a given period T the value of τ determines the distribution of power in the spectrum.

(ii) For a given value of pulse width τ the period T similarly determines the power distribution.

(iii) If we put $T = \tau$ we get a constant (d.c.) level. $|c_n|$ is then given by $A \text{ sinc } n$ so a single spectral line of height A occurs at zero frequency (one might possibly expect this).

(iv) Putting $\tau = T/2$ produces a square wave. Every second harmonic will be zero, as previously found in examples A3.1 and A3.2.

(v) If we let the repetition period T become very large, the line spacing $1/T$ becomes small. As T tends to infinity, the spacing tends to zero and we get a continuous spectrum. Unfortunately, since $|c_n|$ is inversely proportional to T the spectrum vanishes everywhere. This is again because $f(t)$ becomes a finite energy signal if T is infinite, and such signals have zero mean power. However, it is clear that finite energy signals must have continuous spectra.

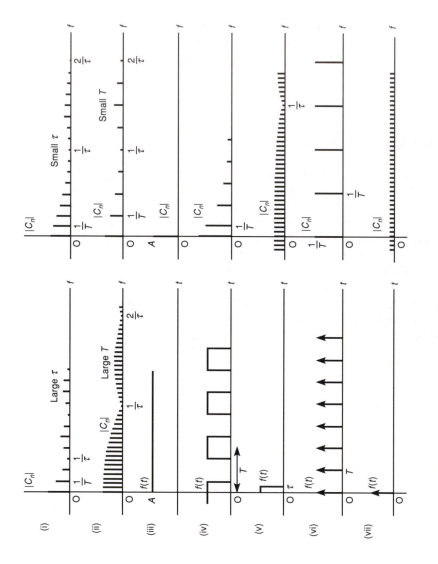

Figure A3.7   *Deductions from the spectrum of Figure A3.5*

(vi) Suppose we make τ small but keep the pulse area $A\tau$ constant. In the limit we get an impulse of strength $A\tau$, and the spectrum will simply be a set of lines of constant height A/T (or $1/T$ for unit impulses).

(vii) Finally, it is clear that a single impulse will have a constant but continuous spectrum, though the same problem as in (v) occurs again.

We have seen previously that a repetitive function can be constructed by convolving a single period of the waveform with a set of impulses spaced at the required repetition period. We will see later that convolution in the time domain is equivalent to multiplication in the frequency domain. The spectrum of the repetitive pulse is thus the product of the spectrum of a single pulse with that of a set of impulses, and therefore has lines spaced at $1/T$ (due to the impulses) with height governed by a sinc function (the shape of the spectrum of a single pulse). These properties are discussed in detail in Chapter 2.

A3.3 Fourier Transforms

We saw above that if we attempt to apply the Fourier Series equations to a finite energy signal, by letting the period T tend to infinity, we can deduce that the spectrum will become continuous since the line separation $1/T$ tends to zero. However, since the expression for c_n contains a $1/T$ outside the integral, c_n also tends to zero.

In order to deal satisfactorily with finite energy signals all that has to be done is to modify equations A3.4 slightly. The line separation $1/T$ is denoted by the differential quantity δf, the (continuous) spectral frequency by $f (= n \times 1/T = n \times \delta f)$, the summation is replaced by an integral and the complex spectrum is defined as $F(f) = \lim_{T\to\infty}(c_n T)$. The Fourier Series relations A3.4 are thus changed to the Fourier Transform relations A3.6.

$$f(t) = \int_{-\infty}^{\infty} F(f)\exp(j\omega t)dt \qquad (A3.6)$$

$$F(f) = \int_{-\infty}^{\infty} f(t)\exp(-j\omega t)dt$$

Although specifically developed here in order to deal with finite energy signals these relations can also be applied to finite power signals, making the Fourier Transform relations particularly powerful. In Chapter 2 these equations are taken as the starting point in the discussion of Fourier Theory and the Fourier Series relations are not used.

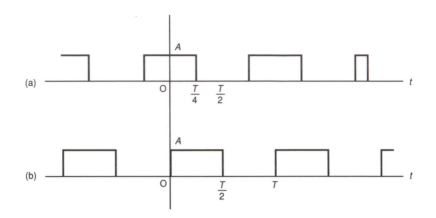

Figure A3.8 *Square waves*

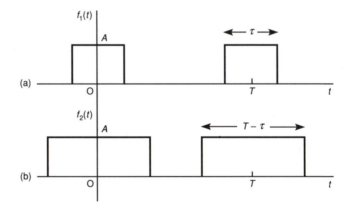

Figure A3.9 *(a) and (b) Repetitive waves*

A3.4 Examples for Solution

1. Use the complex form of Fourier Series to find the coefficients c_n for the square waves of Figure A3.8, taking the origins as shown. Sketch the amplitude and phase spectrum in each case.

2. Show that the complex Fourier coefficients c_n for the repetitive waveform $f_1(t)$ of Figure A3.9(a) are given by: $c_n = \dfrac{A\tau}{T}\operatorname{sinc}\left(\dfrac{n\tau}{T}\right)$

   Show how the repetitive waveform $f_2(t)$ of Figure A3.9(b) can be derived from $f_1(t)$ by simple operations such as addition, inversion and lateral shift, and hence deduce the complex Fourier coefficients for $f_2(t)$.

   Sketch the amplitude and phase spectra for the first eight harmonics of $f_1(t)$ and $f_2(t)$ for $T = 3\tau$.

3.   (i) What is the effect of the zeroth harmonic?

   (ii) Are the even harmonics of a square wave necessarily zero?

   (iii) Are the even harmonics of an even function necessarily zero?

   (iv) What is special about the phase of the harmonics of an even function? And an odd function?

   (v) What is the effect on the harmonics of inverting a function?

   (vi) What is the effect on the harmonics of shifting the origin of a function?

   (vii) How can the harmonics be changed so as to reverse a function (that is, mirror-image it in the y-axis)?

   (viii) If a rectangle wave has a period of 120 units and a pulse width of 20 units, which is the lowest harmonic of zero magnitude? Does this depend on the origin chosen?

   (ix) What is the effect of adding together many harmonics all of the same magnitude and phase?

   (x) How should one select harmonics to produce an approximation to random noise?

Appendix 4: Table of Bessel Functions

Values of Bessel functions $J_n(\beta)$

$n \mid \beta$	1	2	3	4	5	6	7
0	**0.7652**	**0.2239**	−0.2601	−0.3971	−0.1776	**0.1506**	**0.3001**
1	**0.4401**	**0.5767**	**0.3391**	−0.06604	−0.3276	−0.2767	−0.004683
2	**0.1149**	**0.3528**	**0.4861**	**0.3641**	0.04657	−0.2429	−0.3014
3	0.01956	**0.1289**	**0.3091**	**0.4302**	**0.3648**	**0.1148**	−0.1676
4	0.002477	0.03400	**0.1320**	**0.2811**	**0.3912**	**0.3576**	**0.1578**
5		0.00704	0.04303	**0.1321**	**0.2611**	**0.3621**	**0.3479**
6		0.001202	0.01139	0.04909	**0.1310**	**0.2458**	**0.3392**
7			0.002547	0.01518	0.05338	**0.1296**	**0.2236**
8				0.004029	0.01841	0.05653	**0.1280**
9					0.005520	0.02117	0.05892
10					0.001468	0.006964	0.02354
11						0.002048	0.008335

Notes
(1) Values considered significant when determining the number of sidebands for a particular value of β are shown in bold.
(2) A negative sign indicates a phase reversal.

Hence the spectrum for an FM wave, $v = v_c\cos(\omega_c t + \beta\sin\omega_s t)$, with $\beta = 5$, is as shown in Figure A4.1.

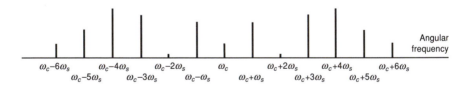

Figure A4.1 *Spectrum of an FM wave*

Solutions to Exercises

Solutions to Exercises on Chapter 1

Exercises 1.1.5

1. (i) The probability of any number is 1/6, so $I = \log_2 6 = 2.58$ bits.
   (ii) The probability of getting a six or a one is $1/6 + 1/6 = 1/3$, so $I = \log_2 3 = 1.58$ bits.
   (iii) Since some number must occur, $p = 1$ and $I = \log_2 1 = 0$.

2. (i) If the three shades are equiprobable, $p = 1/3$ and $I = \log_2 3 = 1.58$ for each dot.
   Therefore, total information $= 20 \times 20 \times 1.58 = 632$ bits
   (ii) The average information per dot is given by

   $$H = -(\tfrac{1}{8} \log 8 + \tfrac{1}{4} \log \tfrac{1}{4} + \tfrac{5}{8} \log \tfrac{5}{8}) = 1.30 \text{ bits}$$

   Therefore, total information $= 400 \times 1.30 = 519$ bits. This is, of course, equivalent to saying that there are 400/8 dots providing $\log 8$ ($= 3$) bits, 400/4 providing $\log 4$ ($= 2$) bits and $(5/8) \times 400$ providing $\log 8/5$ bits. *Note*: If one cannot assume that the arrangement of the dots is known, the information per dot is much greater, each dot having probability $(1/3) \times (1/400)$ in case (i).

3. (i) The average information (per symbol) is

   $$H = -(\tfrac{1}{4} \log 4 + \tfrac{3}{4} \log \tfrac{3}{4}) = 0.81 \text{ bits}$$

   Since the symbols each take 1 s, the rate is 0.81 bits/s.
   (ii) This part cannot be done in the same way. The method to use is to say that in a typical sample of four symbols of the sequence there will be one A and three Bs, and this sample will take 7 s. Each A provides $\log (4) = 2$ bits and each B provides $\log (4/3) = 0.42$ bits. On average in 7 s, one receives $2.0 + 3 \times 0.42$ bits or 0.47 bits/s.
   Note: (i) can be done in the same way, giving one A and three Bs in 4 s.

Exercises 1.2.3

1. $p(M) = p(M, M) + p(S, M)$
   $\qquad = p(M)\, p(M/M) + p(S)\, p(M/S)$
   $\qquad = p(M) \times \tfrac{5}{6} + p(S) \times \tfrac{1}{3}$
   Since $p(M) + p(S) = 1$, we have $p(M) = \tfrac{2}{3}$, $p(S) = \tfrac{1}{3}$.

223

$p(M, M) = p(M).p(M/M) = \frac{2}{3} \times \frac{5}{6} = \frac{5}{9}$
Similarly, $p(S, S) = \frac{2}{9}$, $p(S, M) = p(M, S) = \frac{1}{9}$

$$H(j/i) = -\Sigma\Sigma\, p(i, j) \log p(j/i) \rightarrow 0.74 \text{ bits/symbol.}$$

Redundancy $= (1 - 0.74) = 26$ per cent.

2. $p(B/B) = \frac{1}{2}$, $p(A/B) = \frac{1}{4}$, $p(C/B) = \frac{1}{4}$
   $p(A/A) = \frac{5}{8}$, $p(B/A) = \frac{1}{4}$, $p(C/A) = \frac{1}{8}$
   $p(C/C) = \frac{5}{8}$, $p(B/C) = \frac{1}{4}$, $p(A/C) = \frac{1}{8}$
   Clearly $p(A) = p(C)$ by symmetry

$p(A) = p(A, A) + p(B, A) + p(C, A)$
$\quad\;\; = p(A)\, p(A/A) + p(B)\, p(A/B) + p(C)\, p(A/C)$
$\quad\;\; = \frac{5}{8}\, p(A) + \frac{1}{4}\, p(B) + \frac{1}{8}\, p(C)$

Since $p(A) = p(C)$ and $p(A) + p(B) + p(C) = 1$, we have
$p(A) = p(B) = p(C) = \frac{1}{3}$
$H(j/i) = -\Sigma\Sigma\, p(i, j) \log p(j/i) \rightarrow 1.37 \text{ bits/symbol.}$

Redundancy $= 13.8$ per cent (maximum entropy $= \log_2 3 = 1.58$ bits/symbol).

Note: this solution is very tedious if one does not use the fact that $p(A) = p(C)$. In this example it may be noticed that $p(A, B) = p(B, A)$ etc., but this is not generally true in sequences of more than two symbol types.

3. 1-letter words: a, e (2 words)
   2-letter words: aa, ab, ac, ae, ba, ca, ea, ee, eb, ec, be, ce (12 words)
   3-letter words: bab, bac, cab, cac, beb, bec, ceb, cec (8 words)
   Total number of words $= 22$, Information per word $= \log_2 22 = 4.46$ bits.
   Average length of word $= (\frac{2}{22}) \times 1 + (\frac{12}{22}) \times 2 + (\frac{8}{22}) \times 3 = 2.27$ (3.27 including Ω).
   Average information per letter $= 4.46/3.27 = 1.36$ bits.
   Information per letter if all words independent and equiprobable $= \log_2 5 = 2.32$ bits.
   Redundancy $= 1 - (1.36/2.32) = 0.41$. Probably insufficient for a spoken language, since English has a redundancy of about 80 per cent and is probably the optimum.

4. A A A B A B A A B B A A A A B A A B A A |A

   (a) $p(A) = \frac{14}{20}$, $p(B) = \frac{6}{20}$
       $p(A, A) = \frac{9}{20}$, $p(A, B) = \frac{5}{20}$, $p(B, A) = \frac{5}{20}$, $p(B, B) = 120$
       $p(A/A) = \frac{9}{14}$, $p(A/B) = \frac{5}{6}$, $p(B/B) = \frac{1}{6}$, $p(B/A) = \frac{5}{14}$
   (b) $H(j/i) = -\Sigma\Sigma\, p(i, j) \log p(j/i) \rightarrow 0.854$ bits
       Redundancy $= 1 - 0.854 = 14.6$ per cent.
   (c) Evaluate $p(A/AA)$, $p(A/AB)$, $p(A/BB)$ etc. It should be found that

$p(A/AA) \approx p(A/A)$ etc. if intersymbol effects are only over adjacent symbols (the sequence given is not long enough to verify this). $H(k/ij) = -\Sigma\Sigma\Sigma \, p(i, j, k) \log p(k/ij)$ would have to be evaluated instead of $H(j/i)$.

(d) Certainly not spoken since the redundancy is low.

Exercises 1.3.6

1.

x	A A A A A A	B B B B B B	C C C C C C
y	A A A A B C	B B B B B B	C C C C A B

x	A	A	A	B	C	C	C
y	A	B	C	B	C	A	B
$p(x)$	$\frac{1}{3}$	$\frac{1}{3}$	$\frac{1}{3}$	$\frac{1}{3}$	$\frac{1}{3}$	$\frac{1}{3}$	$\frac{1}{3}$
$p(y)$	$\frac{5}{18}$	$\frac{8}{18}$	$\frac{5}{18}$	$\frac{8}{18}$	$\frac{5}{18}$	$\frac{5}{18}$	$\frac{8}{18}$
$p(xy)$	$\frac{4}{18}$	$\frac{1}{18}$	$\frac{1}{18}$	$\frac{1}{3}$	$\frac{4}{18}$	$\frac{1}{18}$	$\frac{1}{18}$

$I(xy) = \frac{4}{18} \log(\frac{4}{18})/(\frac{1}{3})(\frac{5}{18}) + \text{etc.} \to 0.712$ bits

2. (a) $C = 1 - H(p) = 1 - 0.918 = 0.082$ bits.

(b)

x	A A A A A A A A A A	B B B B B B B B B B
y	A A A A B X X X X	B B B B A X X X X

$I(xy) = 2 \, (\frac{4}{18}) \log \, (\frac{4}{18})/(\frac{1}{2})(\frac{5}{18}) + 2 \, (\frac{1}{18}) \log \, (\frac{1}{18})/(\frac{1}{2})(\frac{5}{18})$

    (AA and BB)   (AB and BA)

$+ \, 2 \, (\frac{4}{18}) \log \, (\frac{4}{18})/(\frac{1}{2})(\frac{8}{18})$

    (AX and XA)

$= 0.154$ bits (note that AX and BX lead to zero information).

(c)

x	A A A A A A A A A	B B B B B B B B B
y	A A A A B X X X X	B B B B A X X X X
obs	A A A A B A A A A	B B B B A A A A A

$I(xy) = \frac{8}{18} \log(\frac{8}{18})/(\frac{1}{2})(\frac{13}{18}) + \frac{4}{18} \log(\frac{4}{18})/(\frac{1}{2})(\frac{5}{18})$

$+ \frac{1}{18} \log(\frac{1}{18})/(\frac{1}{2})(\frac{5}{18}) + \frac{5}{18} \log(\frac{5}{18})/(\frac{1}{2})(\frac{13}{18})$

$= 0.106$ bits.

(d)

x	A A A A A A A A A	B B B B B B B B B
y	A A A A B X X X X	B B B B A X X X X
obs	A A A A B A A B B	B B B B A A A B B

$I(xy) = 2 \, (\frac{6}{18}) \log(\frac{6}{18})/(\frac{1}{2}) \, (\frac{1}{2}) + 2 \, (\frac{3}{18}) \log(\frac{3}{18})/(\frac{1}{2})(\frac{1}{2})$

$= 0.082$ bits.

It is rather surprising that guessing makes things worse, since one intuitively feels that one should be able to do better than chance (or at least as well). However, by comparing the relevant terms it can be seen that although the extra transmissions guessed correctly provide more information, this is more than cancelled by the negative information of those guessed wrongly. This will always be the case if the system has inherently a non-zero information transfer (guessing would not make matters worse only if no information was transmitted at all!).

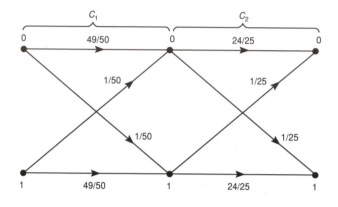

Figure 1.3.6.1 *Cascaded binary symmetric channels*

3. (a) $C_1 = 1 + (p \log p + \bar{p} \log \bar{p}) = 0.86$ bits.
   (b) $C_2 = 0.76$ bits.
   (c) $p(0/0) = (\frac{49}{50})(\frac{24}{25}) + (\frac{1}{50})(\frac{1}{25}) = 0.9416 = \bar{p}$
   $p(0/1) = (\frac{1}{50})(\frac{24}{25}) + (\frac{49}{50})(\frac{1}{25}) = 0.0584 = p$
   $C_3 = 1 - H(p) = 0.679$ bits.

The overall capacity is *always* reduced by cascading noisy channels (no form of 'noise-cancelling' ever occurs).

4. $p(A) = p(AA) + p(BA) + p(CA)$
   $= p(A)\, p(A/A) + p(B)\, p(A/B) + p(C)\, p(A/C)$
   $= (\frac{1}{4})(\frac{2}{3}) + 0$
   $= \frac{1}{6}$

Similarly $p(B) = \frac{1}{2}$, $p(C) = \frac{1}{3}$
$p(AA) = p(A)\, p(A/A) = (\frac{1}{4})(\frac{2}{3}) = \frac{1}{6}$
$p(AB) = {}^{1}/_{12}$, $p(BB) = \frac{1}{4}$, $p(CB) = \frac{1}{6}$, $p(CC) = \frac{1}{3}$, others $= 0$

$I(xy) = \frac{1}{6} \log(\frac{1}{6})/(\frac{1}{4})(\frac{1}{6}) +$ etc. $\rightarrow 0.53$ bits.

Exercises 1.4.5

1. The probability density distribution is shown in Figure 1.4.5.1. It consists of an impulse of strength one-half at the origin plus the usual uniform distribution.

$$H(v) = -\int_{-\infty}^{\infty} p(v)\log p(v)dv = -\frac{1}{2}\log(\frac{1}{2}) + \frac{1}{2}\log(2a)$$

In this case the mean value is $a/4$ and the mean square $a^2/6$.
$H(v) = \frac{1}{2} + \frac{1}{2} \log(24\overline{v^2})^{1/2} = 1.65$ bits compared with $\log(\sqrt{2\pi e})$ bits
($= 2.05$) for the Gaussian.

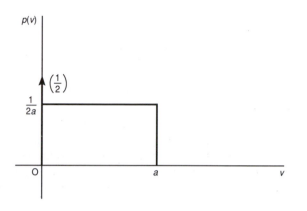

Figure 1.4.5.1 *Probability density distribution of sawtooth waveform*

2. The error rate for binary pulses will be very small with this value of S/N so the information transfer expression will give unity and the information rate is simply given by $2B = 6000$ bits/s. The ideal theorem gives $3000 \log (101) \approx 19\,975$ bits/s.

3. Each pixel will provide $\log(32) = 5$ bits, so the practical information content of the picture is $625 \times 625 \times 5$ bits $= 1.95 \times 10^6$ bits, and the information rate is $1.95 \times 10^6 \times 25 = 48.8 \times 10^6$ bits/s. Ideally the capacity is $8 \times 10^6 \log (1001) \approx 79.7 \times 10^6$ bits/s, which is quite close to the practical rate. Errors in pictorial information are much less important than in numerical data, since the quality of the picture is partly subjective and slight fuzziness etc. has little effect.

Solutions to Exercises on Chapter 2

Exercises 2.2.4

1. (i) The waveform is given by $f(t) = (A/\theta)t$ from $t = 0$ to $t = \theta$.

$$\text{Area} = \int_{-\infty}^{\infty} f(t)\, dt = 2\int_0^\theta \frac{A}{\theta}t\, dt = A\theta$$

$$\text{Energy} = \int_{-\infty}^{\infty} f^2(t)\, dt = 2\int_0^\theta \left(\frac{At}{\theta}\right)^2 dt = \frac{2}{3}A^2\theta$$

(ii) Similarly

$$\bar{f} = \frac{1}{T_0} \int_{-T_0/2}^{T_0/2} f(t) \, dt = \frac{1}{T_0} \int_0^{\theta} f(t) dt = \frac{A\theta}{T_0}$$

$$\overline{f^2} = \frac{2}{3} \frac{A^2 \theta}{T_0}$$

$$\sigma^2 = \overline{f^2} - (\bar{f})^2 = \frac{A^2 \theta}{T_0} \left(\frac{2}{3} - \frac{\theta}{T_0} \right)$$

2. The mean value can be seen to be close to zero. For a very long sequence the mean would always tend to zero in any case. The mean square value is A^2, and is the same as the variance since $\bar{f} = 0$.

Since the waveform is repetitive its power spectrum must contain lines spaced at $1/T_0$. However, the sequence is random, so the power spectrum will tend to be constant (having no dominant frequencies) up to frequencies of the order $1/\theta$; the actual form can be shown to be as in Figure 2.2.4.1.

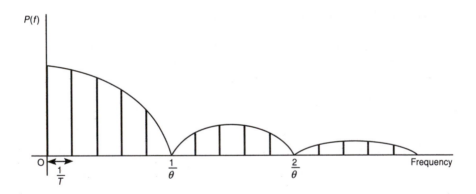

Figure 2.2.4.1 *Power spectrum of pseudo-random sequence*

Such sequences are very important, being used in system testing and in digital noise generators.

Exercises 2.3.6

1. (i) $f_1 \nabla f_1$ is always even, but not necessarily positive (for example if f_1 has both positive and negative sections).

(ii) $f_1 \nabla f_2$ and $f_2 \nabla f_1$ are mirror images.

(iii) $f_1 * f_2$ and $f_2 * f_1$ are identical.

(iv) In general, $f_1 \nabla f_1$ and $f_1 * f_1$ are not the same. They have similar shape if one of the functions is even and are identical if both functions are even.

2.

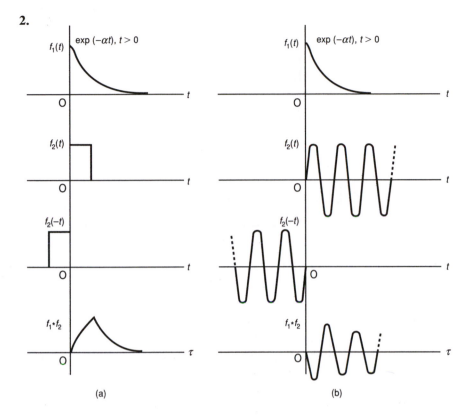

Figure 2.3.6.1 *(a) Convolution of exponential decay with rectangular pulse. (b) Convolution of exponential decay with a burst of sine waves*

These results represent the response of a first-order system to a rectangular pulse or burst of sine waves. If the pulse is very long, the response becomes a step response (exponential rise). The response to a burst of sine waves is a sine wave of the same frequency, but with an initial transient and phase shift. Continuous sine waves would produce a steady amplitude with a constant phase shift.

3.

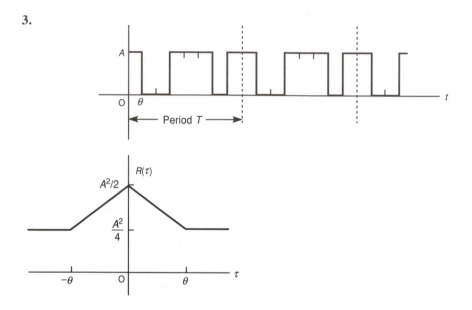

Figure 2.3.6.2 *Pseudo-random sequence and its autocorrelation*

This is a mean power signal; the procedure is unchanged, but the peak value is $A^2/2$ since $R(0) = \bar{f}^2$. $R(\tau)$ falls as τ increases, and is constant for $|\tau| > \theta$ because the pulses are randomly of height A or zero. The probability of a contribution to the integral is then $\frac{1}{2} \times \frac{1}{2}$ giving $A^2/4$ as the constant value.

4. This requires considerably less effort than may be thought at first sight. $f_1 \, \nabla \, f_1$ and $f_2 \, \nabla \, f_2$ are even functions, $f_1 \, \nabla \, f_2$ and $f_2 \, \nabla \, f_1$ are mirror images and $f_1 * f_1$ and $f_2 *f_1$ are identical. The peak values of the functions are important and should be evaluated. Note that although the 'self-convolutions' $f'_1 * f'_1$ and $f_2 * f_2$ are fairly easy to evaluate, they have little practical significance. See Figure 2.3.6.3 on page 231.

Exercises 2.4.9

1. One could use the cosine transform here, since it is easier to integrate cos.cos than cos.exp.

$$F(f) = \int_{-T/4}^{T/4} A\cos(\omega_0 t)\cos(\omega t)\mathrm{d}t$$

This can be integrated by parts, and after a fair amount of rather tedious manipulation gives:

$$F(f) = \frac{AT}{\pi} \frac{\cos(\omega T / 4)}{1 - \omega^2 / \omega_0^2}$$

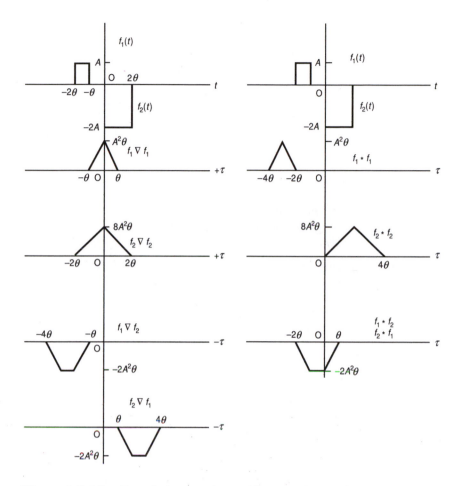

Figure 2.3.6.3 *Correlations and convolutions of two rectangular pulses*

It takes even longer using the full transform. However, the sensible way to proceed is to note that the half-cosine pulse is a product between a continuous cosine wave (spectrum two impulses of strength $A/2$ at f_0 and $-f_0$) and a rectangular pulse of height unity and width $T/2$ (spectrum $T/2 \operatorname{sinc}(fT/2)$). The spectrum is the convolution of these, which is two sinc functions at f_0 and $-f_0$. Therefore

$$F(f) = \frac{AT}{4}\left(\operatorname{sinc}\!\left(fT/2 - \frac{1}{2}\right) + \left(\operatorname{sinc}\!\left(fT/2 + \frac{1}{2}\right)\right)\right)$$

This can be done in one's head, but some manipulation is then required to get it into the form quoted above.

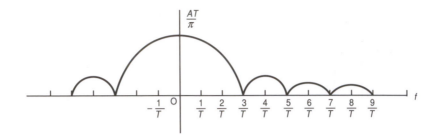

Figure 2.4.9.1 *Amplitude spectrum of half-cosine pulse*

2. Using the addition and shift rules:
$$F(f) = A\tau \, \text{sinc}(f\tau)[\exp(j\omega T) - \exp(-j\omega T)]$$
$$= j4A\tau\pi fT \, \text{sinc}(f\tau) \, \text{sinc}(2fT)$$

  (a) $F(f) = j8A\tau^2\pi f \, \text{sinc}(f\tau) \, \text{sinc}(4f\tau)$
  (b) $F(f) = j2A\tau^2\pi f \, \text{sinc}^2(f\tau)$
  Note that (b) is the case where the two pulses just touch.

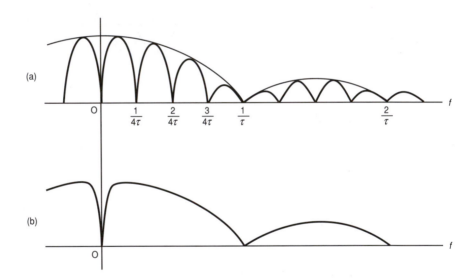

Figure 2.4.9.2 *(a) and (b) Amplitude spectra for double pulse*

3. (i) The spectrum can be found using the differentiation and shift rules. However, it is easier to note that the pulse is the convolution between two rectangular pulses of widths $T/6$ and $T/3$ (the base of the pulse is the sum of the widths, and the sloping side is the width of the narrower pulse). The pulse has area $AT/3$, so

$$F(f) = AT/3 \, \text{sinc}(fT/6) \, \text{sinc}(fT/3) = K \text{ say.}$$

Combining two such pulses, at $T/4$ and $- T/4$

$$F(f) = K\,[\exp(-j\omega T/4) - \exp(+j\omega T/4)] = K2j\,\sin(\omega T/4)$$

Thus $|F(f)| = A\pi f T^2/3\,|\mathrm{sinc}(fT/6)\,\mathrm{sinc}(fT/3)\,\mathrm{sinc}(fT/2)|$

Note that products of sinc functions are very easy to plot.

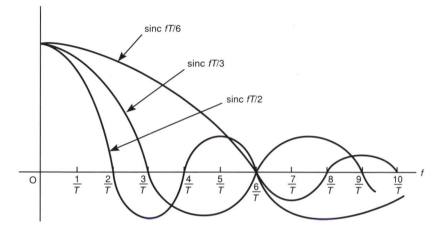

Figure 2.4.9.3 *Amplitude spectrum of trapezoidal pulse*

Repetition at period T produces impulses spaced at $1/T$, but clearly only the first and fifth harmonics will be non-zero. The amplitude ratio of these harmonics is 25.

This example illustrates a good way of producing a relatively pure sine wave by digital means. The choice of pulse shape makes the 2nd, 3rd, 4th, 6th etc. harmonics disappear, so the total harmonic distortion is only about 5 per cent.

4. The triangle pulse has a spectrum $A\theta\,\mathrm{sinc}^2(f\theta)$ and its repetition produces lines spaced at $1/T$, as in Figure 2.4.9.4.

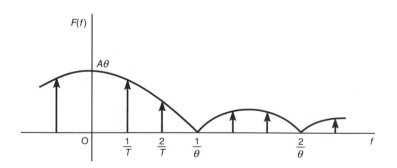

Figure 2.4.9.4 *Amplitude spectrum for $T = 3\theta$*

Clarinet: The even harmonics are zero so $\theta = T/2$. $\text{sinc}^2(fT/2)$ gives lines with ratios 10 : 1. 1: 0.4 etc.

Trumpet: The fourth harmonic is zero so $\theta = T/4$. $\text{sinc}^2(fT/4)$ gives lines with ratios 10: 5: 1.1 etc.

Flute: The second line has to be just below $1/\theta$ to give two lines in the second lobe so $\theta \approx 0.4\,T$. $\text{sinc}^2(0.4fT)$ gives lines with ratios 10: 0.9: 0.4: 0.6 etc.

5. Note that it is not correct to convolve the small triangle with a set of impulses, and then to multiply by the large triangle, as this gives a different waveform of distorted triangles. See Figure 2.4.9.5 on page 235.

$$f(t) = (a(t) \times b(t)) * d(t) \quad F(f) = (A(f) * B(f)) \times D(f)$$

$A(f)$: Impulses of strength $1/a$ at spacing $1/a$
$B(f)$: $4Aa\ \text{sinc}^2\ (4fa)$
$D(f)$: $a/3\ \text{sinc}^2\ (af/3)$

6. $F(t) = A\ \exp(-\alpha t)$
$R(\tau) = A^2/2\alpha\ \exp(-\alpha\,|\tau|)$

$$E(f) = \int_{-\infty}^{\infty} \frac{A^2}{2\alpha} \exp(-\alpha\,|\tau|)\ \cos(\omega\tau)\mathrm{d}\tau \rightarrow \text{(by parts)}\ A^2/(\alpha^2 + \omega^2)$$

$F(f) = A/(\alpha + j\omega)$
$E(f) = |F(f)|^2 = A^2/(\alpha^2 + \omega^2)$

$E = \int f^2(t)\ \mathrm{d}t = A^2/2\alpha$
$E = \int E(f)\ \mathrm{d}f = A^2/2\alpha$

See Figure 2.4.9.6 on page 236.

7. We have previously found $R(\tau)$ for this waveform. The energy spectrum $E(f) = A^2\theta/4\ \text{sinc}^2(f\theta) + A^2/4\ \delta(0)$. The power spectrum $P(f)$ consists of impulses spaced at $1/T$. Ignoring the d.c. component, the power from zero to 1 kHz is $(A^2\theta/4T)$. $2.10^3 \approx 0.16$ V r.m.s. Pseudo-random sequences are often used for generating 'random' noise. A long (maximal length) sequence can be produced by feedback around a shift register, and with a 20-bit register and 1 MHz clock noise essentially white up to, say, 100 kHz is easily produced. In this case the line separation is about 1 Hz so the spectrum is essentially continuous. The method has the advantage that noise of relatively high power is produced with no amplification (and therefore no mains pick-up). See Figure 2.4.9.7 on page 237.

8. I denotes the 'unity' operator and $-I$ the reversing operator.
   (i) $I, I, I\,f, I\,f, -I\,F, -I\,F, I, I$
   (ii) T, T^*, T^*, T

9. This exercise is written in optical spectral code – that is, each symbol is the approximate optical spectrum of an upper-case letter. However, the publishers suspect that one of the authors has used the opportunity of optical cod-

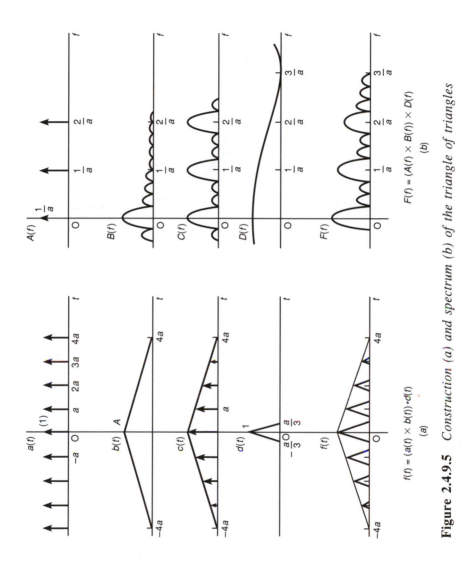

Figure 2.4.9.5   *Construction (a) and spectrum (b) of the triangle of triangles*

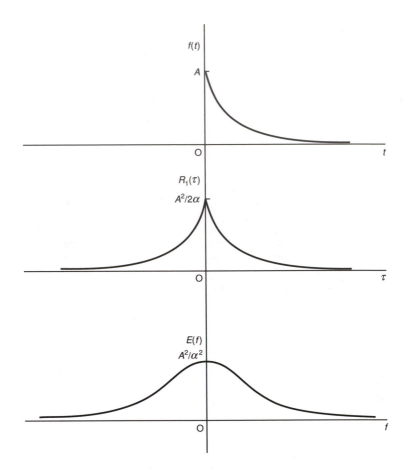

Figure 2.4.9.6 *Autocorrelation and Energy Spectrum for exponential decay*

ing to write an utterly disgraceful and obscene phrase (he refuses to give the solution), and readers are strongly advised not to attempt to decode it.

Exercises 2.5.6

1. $f(t) = \frac{1}{2}\cos(3\omega_0 t) - \cos(\omega_0 t)$

$\omega_0 t$	$\cos(3\omega_0 t)$	$\cos(\omega_0 t)$	$f(t)$
0	$+1$	$+1$	$-\frac{1}{2}$
60°	-1	$+\frac{1}{2}$	-1
120°	$+1$	$-\frac{1}{2}$	$+1$
180°	-1	-1	$+\frac{1}{2}$

See Figure 2.5.6.1 on page 238.

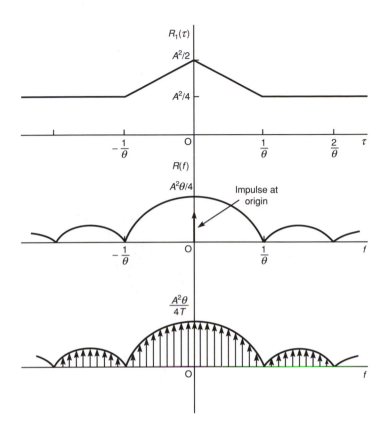

Figure 2.4.9.7 *Autocorrelation and power spectrum for a pseudo-random sequence*

240°	+1	$-\frac{1}{2}$	+1
300°	−1	$+\frac{1}{2}$	−1
360°	+1	+1	$-\frac{1}{2}$

Reconstruction at the ends of the waveform is not complete because the contributions from samples before or after the range chosen is omitted. An infinite number of samples is needed for perfect reconstruction.

2. $f_s(t) = F(t) \times s(t)$ $\qquad$ $F_s(f) = F(f) * S(f)$
$f_{sh}(t) = f_s(t) * h(t)$ $\qquad$ $F_{sh}(f) = F_s(f) \times H(f)$

  (i) Can be recovered by a simple low-pass filter.
  (ii) The distortion due to the hold can be removed by a filter with the inverse characteristic of $H(f)$, that is, $(\mathrm{sinc}\, f\theta)^{-1}$.
 (iii) (a) Recovery is not possible since the sampling rate $< 2B/\mathrm{s}$.
       (b) The hold now has relatively little effect, so recovery is easier.

See Figure 2.5.6.2 on page 239.

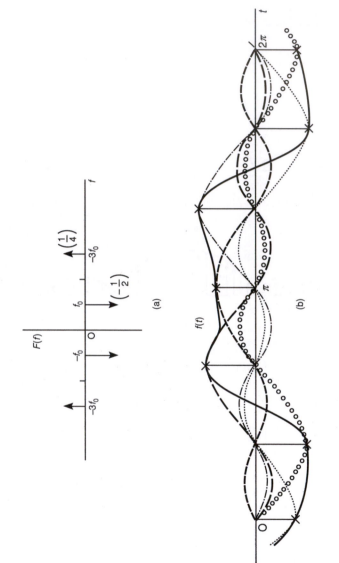

Figure 2.5.6.1 *Recovery of a signal from its samples: (a) spectrum, (b) recovered signal*

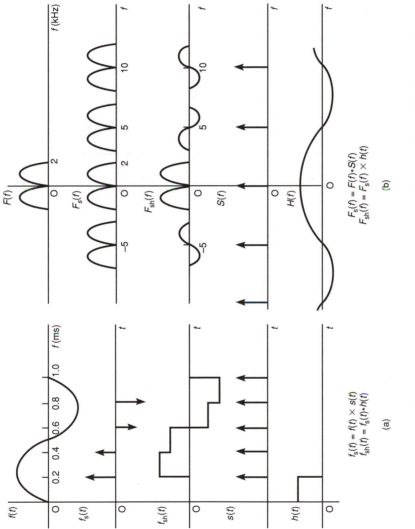

Figure 2.5.6.2 *Signal (a) and spectrum (b) of a sample-and-hold signal*

Solutions to Exercises on Chapter 3

Exercises 3.2.6

1. Using $p_i = (\frac{1}{2})^{l_i}$, $l_i = 1, 2, 3, 4, 4$

s_1	0			
s_2	1	0		
s_3	1	1	0	
s_4	1	1	1	0
s_5	1	1	1	1 (by inspection)

2. Either Fano's or Huffman's code can be used, though Fano's is easier.

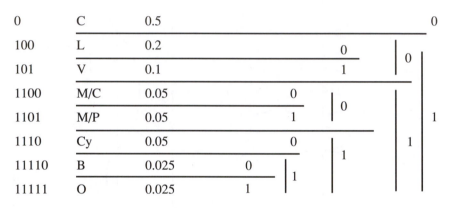

0	C	0.5
100	L	0.2
101	V	0.1
1100	M/C	0.05
1101	M/P	0.05
1110	Cy	0.05
11110	B	0.025
11111	O	0.025

$L = 0.5 + 0.2 \times 3 + 0.1 \times 3 + 0.05 \times 12 + 0.025 \times 10 = 2.25$
$H = -\Sigma p_i \log p_i = 2.21$ bits
$E = 0.98$

Simple binary code: 3 digits needed, $E = 2.21/3 = 0.74$.
Comment: The gain in storage is small and the extra complexity would not be worthwhile.

3. (i) Singly

0	A	0.6 — 0.6 ⌐ 0 1.0
10	B	0.3 ⌐ 0 0.4 — 1
11	C	0.1 — 1

$H = 1.30$, $L = 1.40$, $E = 0.93$

(ii) In pairs

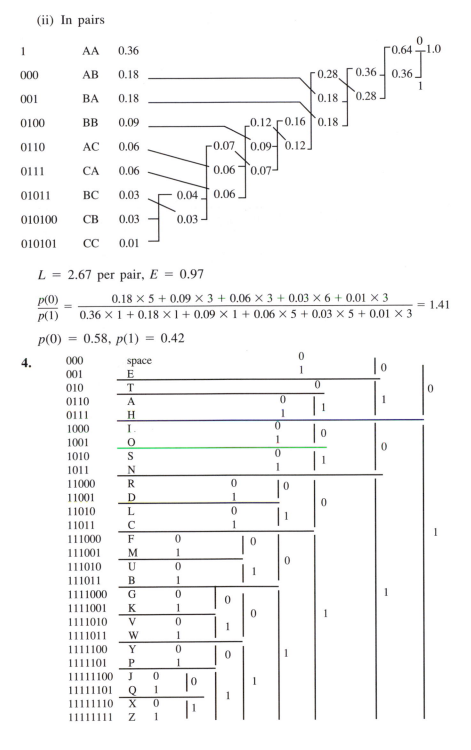

1	AA	0.36
000	AB	0.18
001	BA	0.18
0100	BB	0.09
0110	AC	0.06
0111	CA	0.06
01011	BC	0.03
010100	CB	0.03
010101	CC	0.01

$L = 2.67$ per pair, $E = 0.97$

$$\frac{p(0)}{p(1)} = \frac{0.18 \times 5 + 0.09 \times 3 + 0.06 \times 3 + 0.03 \times 6 + 0.01 \times 3}{0.36 \times 1 + 0.18 \times 1 + 0.09 \times 1 + 0.06 \times 5 + 0.03 \times 5 + 0.01 \times 3} = 1.41$$

$p(0) = 0.58$, $p(1) = 0.42$

4.

000	space
001	E
010	T
0110	A
0111	H
1000	I .
1001	O
1010	S
1011	N
11000	R
11001	D
11010	L
11011	C
111000	F
111001	M
111010	U
111011	B
1111000	G
1111001	K
1111010	V
1111011	W
1111100	Y
1111101	P
11111100	J
11111101	Q
11111110	X
11111111	Z

It should be noted that the procedure is quite easy and quick using the Fano–Shannon method, whereas Huffman's procedure would have required the wall of a large room!

Fortunately no code words are longer than eight digits:

$$L = 3 \times 0.37 + 4 \times 0.36 + 5 \times 0.13 + 6 \times 0.07 + 7 \times 0.06$$
$$+ 8 \times 0.01 = 4.12$$

Saving in memory = 49 per cent (on 8-bits/character). Actually the code is much better than ASCII but only about 20 per cent better than a simple equal-length 5-bit code (though this would lead to worse byte-packing problems).

5.

Incoming character	Pointer output	Added to dictionary
d	4	
a	1	da(28)
b	2	ab(29)
b	2	bb(30)
a	1	ba(31)
da	28	ada(32)
bb	30	abb(33)
ada	32	

So the codes that would be output are 4, 1, 2, 2, 1, 28, 30, 32.

6. The first step is to assign probabilities of occurrence to each character:

a   0.42
b   0.33
d   0.25

Then each character must be assigned an interval within the space 0 to 1:

Character	Low end	High end
a	0	0.42
b	0.42	0.75
d	0.75	1.0

Then the string is encoded:

d	0.75	1.0
a	0.75	0.855
b	0.7941	0.82875
b	0.808653	0.8200875
a	0.808653	0.81345546
d	0.81225485	0.81345546
a	0.81225485	0.8127591014
b	0.8124666214	0.81263303757
b	0.81253653765	0.812591433525
a	0.81253653765	0.8125596046448
d	0.812553822994	0.8125596046448
a	0.812553822994	0.81255626678467

So the information 0.812553822994 would be output from the arithmetic coder. This number could be represented by a binary code of length 40 bits (or five 8-bit characters). The LZW code produced an output of eight 8-bit characters.

Exercises 3.3.4

1. Consider a group of three digits:

$$(\bar{p} + p)^3 = \bar{p}^3 + 3\bar{p}^2 p + 3\bar{p} p^2 + p^3$$
$$\phantom{(\bar{p} + p)^3 = } p(0) \quad p(1) \quad\;\; p(2) \quad\;\; p(3)$$

Single binary errors can be tolerated but two or more produce a symbol error. Therefore $p(\text{group error}) = p(2) + p(3)$, $p_g = 2.8 \times 10^{-2}$.
A symbol error will occur if either or both the two groups comprising a symbol are in error. Therefore

$$(\bar{p}_g + p_g)^2 = \bar{p}_g^2 + 2\bar{p}_g p_g + p_g^2$$
$$\phantom{(\bar{p}_g + p_g)^2 = } p(0)\; p(1) \qquad p(2)$$
$$p(\text{symbol error}) = p(1) + p(2) = 0.055.$$

Note: it is incorrect to use terms of the expansion of $(\bar{p} + p)^6$ since the coding is specifically in threes, not sixes (although 6 digits are used per symbol).

2. (a) Since all combinations are used, one or more binary errors will produce a symbol error.

$$(\bar{p} + p)^5 = \bar{p}^5 + 5\bar{p}^4 p + 10\bar{p}^3 p^2 + \dots$$
$$\phantom{(\bar{p} + p)^5 = } p(0) \quad p(1) \qquad p(2)$$

$p(e) = p(1) + p(2) + p(3) + p(4) + p(5) = 1 - p(0) = 0.226$
Although English is very redundant, an error rate of 1 in 4 to 1 in 5 will produce barely understandable text.

(b) Adding a parity digit means that odd numbers of errors will be detectable, but symbol errors will occur for even numbers of errors.

$$(\bar{p} + p)^6 = \bar{p}^6 + 6\bar{p}^5 p + 15\bar{p}^4 p^2 + 20\bar{p}^3 p^3 + \ldots$$
$$p(0) \quad p(1) \qquad p(2) \qquad p(3)$$

p (parity error) $= p(1) + p(3) + p(5) \approx p(1) = 0.23$
p (symbol error) $= p(2) + p(4) + p(6) \approx p(2) = 0.03$
Symbols may be repeated more than once, until the parity comes out even.
Probability of repetition $= 0.23 + (0.23)^2 + (0.23)^3 + \ldots = 0.30$.
Effective length per symbol $= 6 + 0.3 \times 6 \approx 8$ digits. Therefore effective symbol rate $= 50/8 = 6\frac{1}{4}$ symbols/s. An error rate of 0.03 would produce easily understandable text. It has been commented that this rate is similar to that in a certain well-known daily newspaper.

3. (a) There will be five information digits ($2^5 = 32$) placed in positions 3, 5, 6, 7 and 9. We therefore require parity digits in positions 1, 2, 4 and 8. Position 10 will contain the overall parity digit.

	1	2	3	4	5	6	7	8	9	10
A	0	0	0	0	0	0	0	0	0	0
B	1	0	0	0	0	0	0	1	1	1
C	1	1	0	1	0	0	1	0	0	0
D	0	1	0	1	0	0	1	1	1	1
E	0	1	0	1	0	1	0	0	0	0

Efficiency $= H/L = 0.5$.
(b) $(\bar{p} + p)^{10} = \bar{p}^{10} + 10\bar{p}^9 p + 45\bar{p}^8 p^2 + 120\bar{p}^7 p^3 + \ldots$
$$p(0) \quad p(1) \qquad p(2) \qquad p(3)$$
(i) $p(\text{parity error}) = p(1) + p(3) + p(5) + \ldots \approx p(1) = 0.091$
(ii) $p(2 \text{ or more errors detected}) = p(2) + p(4) + p(6) + \ldots$
$\approx p(2) = 4.1 \times 10^{-3}$
(iii) $p(\text{symbol error}) = p(3) + p(5) + p(7) + p(9) \approx p(3) = 1.1 \times 10^{-4}$
Note: if multiple errors occur it is likely that after attempting to correct the received code word it will be found not to be in the list of allowed code words, so that it will be detected as being in error. Conversely, a large number of even errors could convert one code word into another, so the error would not be detected. The figures above ignore these possibilities, which are difficult to take into account theoretically.

4. $C = 0\ 0\ 1\ 1\ 0\ 0\ 1$
$H^T = 0\ 0\ 1$
$\qquad 0\ 1\ 0$
$\qquad 0\ 1\ 1$
$\qquad 1\ 0\ 0$
$\qquad 1\ 0\ 1$
$\qquad 1\ 1\ 0$
$\qquad 1\ 1\ 1$

The multiplication is performed modulo-2 and the result can be seen to be all zeros.

5. The CRC generator polynomial is 1101. This must be divided into each of the message sequences (modulo-2). If the division yields an exact result then the data are correct and consists of the whole sequence, less the least significant 3 bits (the remainder). The first sequence is in error. The second is correct and the data sequence is 11011. The third sequence is correct and the data sequence is 1011011.

Solutions to Exercises on Chapter 4

Exercises 4.5

1. (i) The mean square value of the modulated wave is $\overline{v_m^2} = \frac{1}{2} a_c^2 + \frac{1}{4} m^2 a_c^2$, so the power increases as m^2. However, even for $m = 1$ the sidebands carry only one-third of the total power.

   (ii) Each sideband contains all the information, but AM as opposed to SSAM is normally used as it is much simpler to implement (SSAM does not include the carrier in its spectrum, and one must be generated at the receiver).

   (iii) An FM wave has a constant amplitude independent of β, so the power is constant.

   (iv) The amplitude spectra are very similar with β very small. However, the phase spectra are different. Narrow-band FM is used for some special applications, since it still has better noise tolerance than AM though its power advantage is lost.

   (v) AM has advantages of narrow bandwidth and simple implementation. FM requires a larger bandwidth and a more complex receiver, but requires much less transmitted power for a given output-power-to-noise ratio (that is, its tolerance of noise is much better).

2. (i) Assume that the speech bandwidth is 3 kHz and sample at 8000/s. If we digitise into 8 levels (3 bits) the information will be 24 000 bits/s. The recording time will be $10^6/24.10^3 \approx 40$ s, so the system would be rather limited in its application. With only 8 levels the quantisation noise will be high, with $(S/N)_o \approx 64$.

   (ii) The outer track has length $\approx 2\pi \times 50$ mm and carries $(2\pi.50.10^{-3})/(2.10^{-6}) \approx 16 \times 10^4$ bits, so the 'mean' track will carry about 8×10^4 bits. There will be about $(50 \times 10^{-3})/(10.10)^{-6} = 5000$ tracks, so the total number of bits $\approx 4 \times 10^8$.

   For high-quality music we need a bandwidth of at least 15 kHz, so the sampling rate $\approx 30\,000$/s. A high root-mean-square signal-to-noise ratio is needed, say 60 dB at least, equivalent to $(S/N) \approx 10^6$, so the number of levels required is $M = 10^3$. Ten bits are therefore needed per sample, so the bit rate is 3×10^5/s.

The playing time will be $(8 \times 10^8)/(3 \times 10^5) \approx 3000$ s or 50 minutes, so the system is clearly very feasible. The speed of rotation of the disc can be deduced, since there are 5000 tracks, and is 5/3 rev/s or 100 rev/min.

3. (i) There are 8 tones and one rest, so the information is log 9 = 3.2 bits per tone. The average length per tone is 3/4 s giving 4.27 bits/s or 21.1 bits in 5 s. If the number of 'tunes' is N, then assuming they are equiprobable $\log N = 21.1$ so $N = 2^{21.1} \approx 2.2 \times 10^6$. Quite a lot of 'tunes', but many of them would not be very interesting!

    (ii) The spectrum of a given tone lasting a given time is a sinc function (an impulse for a continuous tone convolved with a sinc function for the duration) but since the shortest duration is 0.2 s (= 5 Hz) this spreading effect is small, and the bandwidth can be taken as 512 − 256 = 256 Hz.

    (a) Double sideband AM can easily be used, since the channel bandwidth will be about 500 Hz, so a carrier at about 1.5 kHz could be chosen.

    (b) With $\beta = 5$, $B \approx 2(\beta + 1) \times 256 = 3$ kHz, so FM is just possible.

    (c) With PCM we could sample at say 3×256/s and with 256 levels we have $m = 8$ so the channel bandwidth will be $32 \times 8 \times 256 \approx 3$ kHz. This would give a good power-to-noise ratio.

    (d) Each tone requires log 9 = 3.2 bits (4 needed in practice) for pitch plus log 4 = 2 bits for tone. Adding one bit for synchronisation the maximum rate is 7 bits in 0.2 s $\approx$ 35 bits/s. This requires a bandwidth of only about 20 Hz. The system complexity is higher, of course, since the tones have to be regenerated from the code received.

4. The waveforms for the data 100110000010111 when encoded using differential-Manchester and HDB3 coding are shown in Figure 4.5.1:

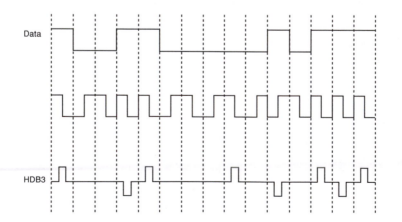

Figure 4.5.1

To transmit using QAM it is necessary to divide the data into 4-bit blocks. As the number of bits is not exactly divisible by 4, an extra zero is put at the start of the data. Hence, the blocks are: 0100 1100 0001 0111. Using the QAM diagram these translate to phase/amplitude changes of:

0100	+157.5	L
1100	+202.5	L
0001	+45	L
0111	+135	L

This would look something like Figure 4.5.2.

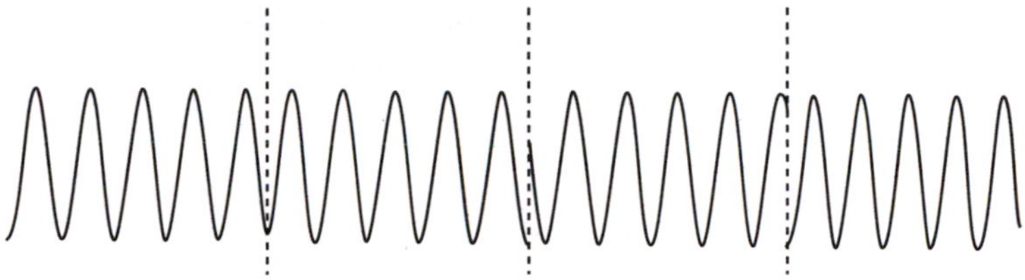

Figure 4.5.2 *Phase transition intervals*

Solutions to Exercises on Chapter 5

Exercises 5.5

1. Suppose the pad measures 10 cm × 10 cm. Using an 8-bit ADC gives a resolution of $100/256$ mm ≈ 0.5 mm, probably enough for most applications. In general, for a square of side L an n-bit ADC will give resolution of $L/2^n$, producing $2n$ bits/sample. With a 3 kHz channel the number of samples/s $= 3000/2n$ and the time to traverse the page is $2^n/(\text{samples/s})$ $= 2^n \times n/1500$

n	Resolution	Time/line (s)
4	6.25	0.04
6	1.56	0.26
8	0.39	1.37
10	0.10	6.83
12	0.02	32.8

8-bits is clearly a reasonable compromise.
The particular advantage over a facsimile system is that only lines drawn are 'scanned', rather than every possible line (many of which are blank). The proposed system is clearly good for sketches, circuit diagrams, signatures etc., but its advantages fall as the complexity of the sketch etc. increases.

2.   (i) There are $400 \times 600 = 24 \times 10^4$ 'pixels' per page, and the maximum bit rate is 3000/s (using AM with 1.5 kHz per sideband). The time taken to transmit a page is $24 \times 10^4/3000 \approx 80$ s. Slow, but can handle documents, pictures etc.

   (ii) Assuming five letters per word on average, number of bits/page $= 300 \times 5 \times 5$. The time taken is $300 \times 5 \times 5/3000 \approx 2.5$ s. Fast, but expensive hardware.

  (iii) Speaking at 2 words/s takes about 150 s. Rather a waste of time and money, unless speaking to one's girlfriend. Ideally $C = B\log(1 + S/N) \approx 3 \times 10^4$ bits/s, or about ten times faster.

3. (a) The theoretical capacity is the maximum error-free communication rate given by $C = B\log(1 + S/N)$. The source rate is the actual rate, ignoring errors.

     When sampling a signal the number of distinguishable levels is limited to about $\sqrt{S/N}$ and with $2B$ independent samples/s this gives $2B\log(S/N)^{\frac{1}{2}} = B\log(S/N)$.

 (b) (i) Teletext system

       Assume $B = 8$ MHz, $S/N = 10^3$ (32 levels). Then
$C_{\text{ideal}} = 8 \times 10^6 \log(10^3) \approx 8 \times 10^7$.
Teletext uses 4/625 lines so $C_{\text{ideal}} \approx 0.51 \times 10^6$ bits/s.
The data rate is 45 bytes in 52 μs $= 6.9 \times 10^6$ bits/s. One page takes 6/25 s (4 lines in 1/25 s). Data per page $= 24 \times 40 \times 7 = 6720$ bits, so actual data rate ≈ 2800 bits/s and $R/C \approx 0.055$.
Since $S/N \approx 10^3$ there will be several errors per page, but a single parity bit is sufficient and leads to blanking of characters. Page errors would be more important and an (8, 4) Hamming code (single-error-correcting, double-error-detecting) is used for headers.

    (ii) FAX system

       Assume $B = 3$ kHz and $S/N = 100$. Ideally $B = 3000\log(10) \approx 2 \times 10^4$ bits/s.
Data per page $= 1188 \times 1728 \approx 2.05 \times 10^6$ bits. A telephone line accepting $2B = 6000$ bits/s would therefore take an unacceptably long time per page (≈ 5 minutes). Run-length encoding is used, followed by modified Huffman coding, producing a compression of about 30. $R/C \approx \frac{1}{3}$ (6000/20 000).
There is no error correction in FAX, though it is often pictorial so errors are less important than in Teletext.

4. (a) Block codes consist of n digits per block, of which k carry information. Hamming Distance: number of digit positions in which two codewords differ.
Minimum Hamming Distance d_{min} : least difference between any two codewords in complete set.

Error detection: $d_{min} = E + 1$ for E errors
Error correction: $d_{min} = 2E + 1$ for E errors.
(b) Shannon's theorem states that the maximum error-free rate is given
by $C = B \log(1 + S/N)$. A typical system operates at a rate $R < C$
with a finite error rate.
(i) Teletext
See the details in question 3 above. $C = 8 \times 10^6 \log(10^3) \approx 8 \times 10^7$
bits/s. $R = 6.9 \times 10^6$ bits/s. Error rate is typically 1 or 2 per page, say
1 in 10^4, $R/C \approx 0.1$ ($6.9 \times 10^6/8 \times 10^7$).
(ii) CD system (music)
A very low error rate is needed since bit errors in significant positions
would be very serious. The system uses two stages of (32, 28) and (28,
24) cross-interleaved Reed–Solomon coding (cyclic block code using bytes)
leading to almost negligible error rates (for reasonably low binary error
rates, say 10^{-3}). On decoding the first stage corrects single byte (random)
errors and flags multiple errors and the second stage corrects multiple
errors (or interpolates or mutes if not possible).
The data comprise two 16-bit samples at 44.1 kHz, giving $R =$
176 kbytes/s $= 1.4 \times 10^6$ bits/s. The pit size is about 1 μm and scanning
speed 1 m/s, so the equivalent bandwidth is about 500 kHz, corresponding to $R = 1.4 \times 10^6$ bits/s.
$S/N \approx M^2 = 2^{16}$ so $S/N \approx 2^{32}$.
$C = 500 \times 10^3 \log(2^{32}) \approx 16 \times 10^6$ bits/s, so $R/C \approx 0.1$.

Solutions to Exercises on Appendix 1

Exercises A1.4

1. (i) A probability density distribution $p(x)$ has the property that $\int p(x) \, dx = 1$.
    Therefore $\int a \exp(- |x|/b) dx = 2ab = 1$ so $a = 1/2ab$.
    $\bar{x} = 0$ since $x\, p(x)$ is odd and the integral must be zero.
    $$\bar{x}^2 = \sigma^2 = \int_{-\infty}^{\infty} x^2 \, p(x) dx = 2 \int_0^{\infty} \frac{x^2}{2b} \exp(- |x|/b) dx = 2b^2$$
    (by parts)
    (ii) Proceeding as before:
    $$p(\text{error}) = \int_v^{\infty} p(x) dx = \frac{1}{2b} \int_v^{\infty} \exp(-x/b) dx$$
    $$= (1/2) \exp(- |v|/b)$$
Note that $p(\text{error}) = \frac{1}{2}$ for zero signal, as may be expected.
Note: noise normally has a Gaussian distribution, of course, but an exponential one is used here since it can be integrated. If one wants to set a
question using a Gaussian one has to include the appropriate tables of

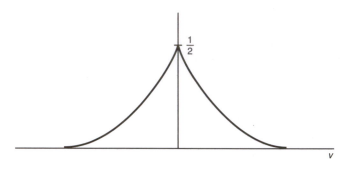

Figure A1.4.1 *Error probability distribution*

∫ exp $(-t^2/2)$ dt etc., which is a considerable nuisance. The writer considers this to be yet another example of the inherently unfair nature of the world.

2. The mean modulus η of the sine wave of Figure A1.4.2(a) is

$$\eta = \frac{1}{T/2} \int_0^{T/2} a\ \sin(\omega t)\mathrm{d}t = \frac{2a}{\pi}$$

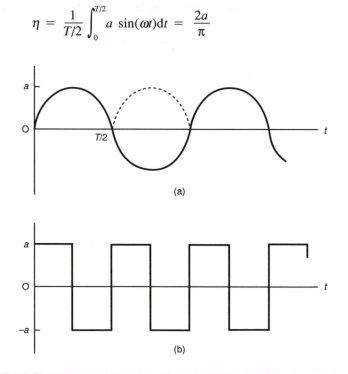

Figure A1.4.2 *Mean modulus of (a) sine wave, (b) square wave*

The mean square value $(= \sigma^2) = a^2/2$ so $\sigma = \pi\eta/(2\sqrt{2}) = 1.11\eta$.
The calibrated scale reading is therefore $1.11 \times \eta$ (the meter 'measures' η).
  (i) The square wave of Figure A1.4.2(b) has $\eta = a = \sigma$. The scale
      reading is therefore 1.11η, and the meter reads 11 per cent high.
 (ii) The mean modulus of Gaussian noise is

$$\eta = 2 \int_0^\infty x\, p(x)\mathrm{d}x = \frac{2}{\sigma\sqrt{2\pi}} \int_0^\infty x\exp(-x^2/2\sigma^2)\ \mathrm{d}x$$

   This can be integrated easily, giving $\eta = \sigma/(\sqrt{\pi/2})$.
   The scale reading is therefore $1.11\eta = 0.89\sigma$.
   It therefore reads 11 per cent low.

3. (a) The probability density can be deduced intuitively for a square wave;
       there must be indefinitely narrow spikes (impulses) at the two possi-
       ble amplitude values, and each spike must have 'area' of 0.5 unit.
       Similarly, a triangle wave has a uniform distribution. For a sine wave
       $x = a \sin \omega t$ and $p(x) = 1/(\text{period} \times \text{slope}) = 1/(a\pi\cos \omega t) =$
       $1/(\pi(a^2 - x^2)^{1/2})$. The functions and corresponding probability densi-
       ties are shown in Figure A1.4.3.

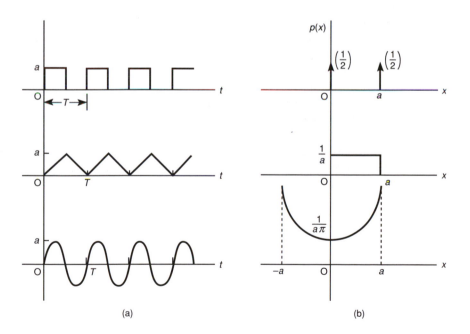

(a)   (b)

Figure A1.4.3 *Various waveforms (a) and their probability density
distributions (b)*

(b) (i) Ensemble averages $\int x^2 \, p(x)dx$
Square wave: $a^2/2$
Triangle wave: $a^2/3$
Sine wave:

$$\int_{-a}^{a} \frac{x^2}{\pi(a^2 - x^2)^{\frac{1}{2}}} \, dx \rightarrow a^2 / 2$$

(ii) Time averages $\dfrac{1}{T_0} \displaystyle\int_{-T_0/2}^{T_0/2} x^2(t)dt$

Square wave: $a^2 /2$
Triangle wave: $a^2/3$
Sine wave: $a^2 /2$

Solutions to Exercises on Appendix 2

Exercises A2.4

1. (i) From the Principle of Equipartition, the energy of the electrons (and their mean-square velocity) is directly proportional to absolute temperature, so the total noise is proportionally increased. There would be no noise at absolute zero.

   (ii) Johnson noise is caused by current pulses due to collisions of electrons with the lattice. A direct current would have no effect (unless the resistor was significantly heated). Johnson noise does occur in transistors, owing to their ohmic base resistance.

   (iii) Shot noise depends essentially on current. It depends on temperature only insofar as temperature may change the current. It does not occur in a resistor carrying a direct current, since there is no potential barrier determining the current.

2. Referring to Figure A2.5 the effective resistance of the diode is only 250 Ω, so only a fraction $(R_d/(R_d + R))^2 \approx R_d^2/R^2$ of the previous voltage is measured at the output.

$$\overline{v_0^2} = (4RkT\Delta f + 2eIR^2\Delta f) \times R_d^2/R^2$$
$$= 2.1 \times 10^{-12} \text{ V}^2 \text{ (about 1.4 μV r.m.s.)}$$

If the current is increased to 1 mA the Shot noise rises by a factor of 10 but the semiconductor diode resistance falls by the same amount, the Johnson noise remaining the same.

   (i) Thermionic diode

$$\overline{v_0^2} = (1.6 \times 10^{-10}) + (3.2 \times 10^{-8})$$
$$= 3.2 \times 10^{-8} \text{ V}^2, \text{ an increase of nearly a factor of 10.}$$

(ii) Semiconductor diode

$$\overline{v_o^2} = (V_J^2 + V_S^2) \times \frac{R_d^2}{R^2} \approx 2.1 \times 10^{-13} \text{ V}^2.$$

The observed noise actually falls by about a factor of 10; in practice the noise of a following amplifier would be dominant and little change would be seen. In practice it is very difficult to demonstrate the Shot noise formula in semiconductor diodes, since an amplifier of very low input impedance and very low noise is required, whereas it can be demonstrated very easily with thermionic diodes. However, one usually produces noise either by a pseudo-random sequence or by using a Zener diode in the breakdown region these days.

Solutions to Exercises on Appendix 3

Exercises A3.4

1. (a) $c_n = \dfrac{1}{T} \displaystyle\int_{-T/4}^{T/4} A \exp(-jn\omega_0 t)dt = \dfrac{A}{n\pi} \sin(n\pi/2) = \dfrac{A}{2} \operatorname{sinc}(n/2)$

(b) $c_n = \dfrac{1}{T} \displaystyle\int_{0}^{T/2} A \exp(-jn\omega_0 t)dt = \dfrac{A}{n\pi} \exp(-jn\pi/2) \sin(n/\pi 2)$

$= \dfrac{A}{2} \operatorname{sinc}(n/2) \exp(-jn\pi/2)$

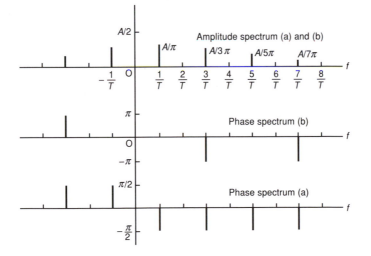

Figure A3.4.1 *Amplitude and phase spectra for square waves*

2. $c_n = \dfrac{A\tau}{T} \, \text{sinc}(n\tau/T)$

The amplitude spectra are the same, except for an extra line of height A for $f_2(t)$, and are shown in Figure A3.4.2.

3.   (i) The zeroth harmonic gives the mean (d.c.) level.

(ii) Yes. In a perfect square wave they are exactly zero.

(iii) No, for example for repetitive pulse.

(iv) The phases of the harmonics of an even function are all zero or $\pm\pi$ (because they refer to cosine waves). Similarly those of an odd function are all $\pm\pi/2$ (they refer to sine waves).

(v) The phases are changed by $\pm\pi$ (that is, the harmonics are inverted).

(vi) A phase term $\exp(-jn\omega_0\tau)$ is produced for a shift of τ. This moves the origin of the first harmonic by τ, but higher harmonics have to be shifted by greater amounts to keep in step.

(vii) The phases have to be reversed.

(viii) The sixth (120/20), and it does not depend on the origin.

(ix) A set of impulses is produced

(x) Add all the harmonics with the same amplitude but random phase.

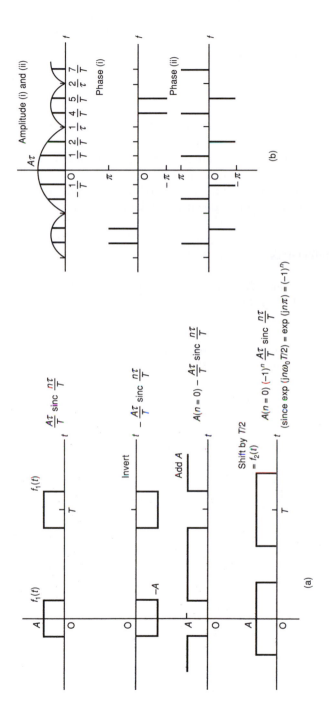

Figure A3.4.2 *Pulse waveforms (a) and their amplitude spectra (b)*

Classic References and Bibliography

The literature of Information Theory and of Communications is very extensive. However, the vast numbers of references that many authors delight in are often somewhat overwhelming and relatively unhelpful in getting into a new subject and the references quoted here have been kept to a minimum. A short Bibliography is also included of books providing valuable background information or more detail as explained below.

Classic References

Bose, R., Chandhuri, D. and Hocquenghem, A. (1960). *Int. J. Information Control*, Vol. 3, p. 68.

Dudley, H. (1936). 'Synthesising Speech', *Bell Laboratories Record*, Vol. 15, p. 98.

Fano, R.M. (1963). *IEF, E Trans. Information Theory*, Vol. IT9, p. 64.

Hagelburger, D.W. (1959). *Bell System Technical Journal*, Vol. 38, p. 969.

Hamming, R.W. (1950). *Bell System Technical Journal*, Vol. 29, p. 147.

Hartley, R. (1928). *Bell System Technical Journal*, Vol. 7, p. 535.

Huffman, D.A. (1952). *Proc. IRE*, Vol. 40, p. 1098.

Johnson, J.B. (1928). *Phys. Rev.*, Vol. 32, p. 97.

Nyquist, H. (1924). *Bell System Technical Journal*, Vol. 3, p. 324.

Nyquist, H. (1928). *Phys. Rev.*, Vol. 32, p. 110.

Rice, S.O. (1950). *Bell System Technical Journal*, Vol. 29, p. 60.

Shannon, C.E. and Weaver, W. (1949). *A Mathematical Theory of Communication*, Univ. of Illinois Press, Champaign, Illinois.

References quoted in the text and not listed above

Abramson, N. (1963). *Information Theory and Coding*, McGraw-Hill, London and New York.

Bell, D.A. (1953). *Information Theory*, Pitman, London and New York.

Fano, R.M. (1961). *Transmission of Information*, MIT Press, Cambridge, Massachusetts.

Howatson, A.M., Lund, P.G. and Todd, J.D. (1972). *Engineering Tables and Data*, Chapman & Hall, London.

Rosie, A.M. (1973). *Information and Communication Theory*, 2nd edn, Van Nostrand Reinhold, New York.

Stuart, R.D. (1966). *An Introduction to Fourier Analysis*, Science Paperbacks, Methuen, London.

Bibliography

N. Abramson, *Information Theory and Coding*, McGraw-Hill, New York, 1963. Comprehensive treatment of principles of coding in discrete systems.

J. Brown and E.V. Glazier, *Telecommunications*, Science Paperbacks, Methuen, London, 2nd edn, 1974. Deals primarily with telecommunications but provides some useful information on Information Theory.

J.P. Cater, *Computer Speech Generation*, Howards Sams, Indianapolis, Indiana, 1983. Primarily for laymen but gives useful details on available systems.

F.R. Connor, *Signals*, Edward Arnold, London, 2nd edn, 1983. Useful introduction to signal analysis, coding and Information Theory.

J. Dunlop and D.G. Smith, *Telecommunications Engineering*, Chapman & Hall, London, 3rd edn, 1994. A good introduction to most aspects of telecommunications, without too much maths.

M.A. Jack (editor), *The Impact of Microelectronics Technology*, Edinburgh University Press, Edinburgh, 1982. Deals with applications of Information Technology, including Teletext and Viewdata.

M. Nelson, *The Data Compression Book*, M&T Books, New York, 2nd edn, 1995. Still the most comprehensive introduction to data compression.

J.G. Proakis, *Digital Communications*, McGraw-Hill, New York, 3rd edn, 1995. A highly mathematical and comprehensive treatment of all aspects of digital communications.

A.M. Rosie, *Information and Communication Theory*, Gordon and Breach, New York, 1987. Comprehensive coverage of many aspects of Information and Communication Theory.

M. Schwartz, *Information Transmission, Modulation and Noise*, McGraw-Hill, New York, 4th edn, 1990. Very thorough treatment of communication aspects of signal theory.

Bernard Sklar, *Digital Communications*, Prentice-Hall, Hemel Hempstead, 1988.

R.D. Stuart, *An Introduction to Fourier Analysis*, Science Paperbacks, Methuen, London, 1966. Relatively readable, though mathematical, introduction to Fourier Series and Transforms.

A.S. Tanenbaum, *Computer Networks*, Prentice-Hall, Hemel Hempstead, 3rd edn, 1996. An excellent coverage of all aspects of computer communications.

P. Woodward, *Probability and Information Theory with Applications to Radar*, Pergamon, New York, 2nd edn, 1964. Compact and elegant approach to signal and Information Theory, but requires some prior knowledge.

Index